P9-AFT-488

A
FIRST GREEK COURSE

by

W. L. DONALDSON, M.A.

Late Scholar of Worcester College, Oxford, and
Assistant Master at Haileybury College

CAMBRIDGE
AT THE UNIVERSITY PRESS
1970

Published by the Syndics of the Cambridge University Press
Bentley House, 200 Euston Road, London, N.W.1
American Branch: 32 East 57th Street, New York, N.Y. 10022

ISBN 0 521 04851 6

First Edition 1931
Reprinted 1937 1946 1949 1952
1956 1960 1964 1970

Printed in Great Britain
at the University Printing House, Cambridge
(Brooke Crutchley, University Printer)

PREFACE

THERE is no royal road to the acquirement of Greek. Of that the author of this book is well aware. Yet he believes that the road can be shortened, that it can be made more interesting. This book attempts to combine grammar, reading and composition.

Boys as a rule nowadays begin Greek late. They have already a knowledge of grammar in general, through Latin, or through the French verb, and do not need the minute instructions given in most books. They are expected to reach a good standard more quickly.

The Dual is omitted at first. Anyone who has reached p. 88 can learn it all in ten minutes, but there is nothing to prevent the teacher from administering it in small doses if he thinks fit.

Particles are introduced with the first sentences. A page of Greek without particles is not Greek at all.

The Infinitive is introduced with each tense that is learnt and simple usages that cause no difficulty are put in practice as early as possible.

Participles. As soon as the declension of these is known they are used as freely as possible.

Accents. Opinion varies so much as to the value of teaching the accents and pronunciation from the first that these two points have been left to the individual teacher.

The Greek examples have, so far as is possible, been selected with some time and trouble from good Greek authors, with occasional shortening and adaptation. The story on p. 55 "*With Brains, sir!*" was taken from an old-fashioned but excellent reader by W. G. Rushbrooke, and the author of the present book is doubtful whether it is original Greek or not.

W. L. D.

July 1931

SUGGESTIONS FOR THE TEACHER

(1) It is recommended that plenty of time should be given to teaching the Alphabet by writing each letter (capital, name and script) on the board while the pupils practise writing them thus:

α α α α Α
β β α α β α β Α Β
γ γ β α γ α β γ Γ Β Γ etc.

The teacher meanwhile can be giving a commentary on their origin or, perhaps, the reasons for the different order of the Latin Alphabet.

(2) The list of English words which are also Greek words has been found to attract the interest of beginners while providing practice in writing, and should be done carefully. "Half the English Dictionary is Greek."

(3) A book of Greek Tales in English, such as Kingsley's *Heroes* and, later, a History of Ancient Times linking up Egypt, Babylonia, Crete and Persia with the History of Greece might be read at intervals, not necessarily as a lesson for marks.

(4) Certain pages at the end of the book might be studied, in part or in whole, at suitable stages, *e.g.* p. 97 I and II after p. 45. The first three pieces on p. 100 contain no Subjunctive nor Optative.

(5) A Greek Grammar should be taken into use as soon as the Active of the Verb is known. Though the present book is, in a way, self-sufficing there are some boys who profit by having the whole verb on one page and will learn irregular verbs on the sly in order to surprise you!

(6) If time permits, an easy reader, such as *Tales from Herodotus*, should be brought into use after learning the Subjunctive.

CONTENTS

PART I

PART II

PART III

PART I

THE ALPHABET

Α	α	Alpha	= a
Β	β	Bēta	= b
Γ	γ	Gamma	= g (hard, as in 'gun')
Δ	δ	Dĕlta	= d
Ε	ε	Ĕ-psilŏn	= ĕ (as in 'met') thin ĕ
Ζ	ζ	Zēta	= z (= δσ)
Η	η	Ēta	= ē (as in 'meet')
Θ	θ	Thēta	= th (as in 'thin')
Ι	ι	Iōta	= i
Κ	κ	Kappa	= k
Λ	λ	Lambda	= l
Μ	μ	Mu	= m
Ν	ν	Nu	= n
Ξ	ξ	Xi	= x (κσ, γσ, χσ)
Ο	ο	O-mīcrŏn	= ŏ (as in 'got') o-little
Π	π	Pi	= p
Ρ	ρ	Rhō	= r (rh)
Σ	σ ς	Sigma	= s (ς only at the end) never = **z**
Τ	τ	Tau	= t
Υ	υ	U-psilon	= u (thin u)
Φ	φ	Phi	= ph
Χ	χ	Chi	= ch, kh (as in 'chorus')
Ψ	ψ	Psi	= ps (βσ or πσ)
Ω	ω	O-mega	= ō (as in gō) o-big

γ takes the place of ν before γ, κ, χ and ξ, *e.g.* 'larynx' = λάρυγξ, 'sponge' = σπόγγος.

The aspirate (h), at the beginning of a word only, is written thus: ἁ = ha. Initial ρ and υ always have the aspirate.

This sign reversed is placed over unaspirated vowels which begin a word: ἐξ. These signs stand to the left of a capital letter: Anchises Ἀγχίσης.

When a diphthong (δίφθογγος) begins a word the sign is placed over the second vowel: *αὐτό, εὕρηκα.*

ττ is preferred to σσ in Attic Greek.

CLASSIFICATION OF CONSONANTS AND VOWELS

Vowels are classified as: Hard α ε ο Soft ι υ
η ω

Diphthongs: αι ει οι υι } ear diphthongs.
αυ ευ (ηυ) ου }
ᾳ ῃ ῳ eye diphthongs (ι not sounded).

This is called iota subscriptum or subscript.

Consonants: (1) *Mutes*

	Hard	Soft	Aspirated	Nasal
Guttural = Throat sounds	κ	γ	χ	γ = (νγ)
Labial = Lip sounds	π	β	φ	μ
Dental = Tooth sounds	τ	δ	θ	ν

(2) *Semi-vowels*

Liquid λ ρ
Nasal μ ν
Sibilant σ (ς)

Double Consonants: Guttural sound + σ = ξ
Labial sound + σ = ψ
Dental sound + σ = ζ

Stops: Full-stop and comma as in English.
Colon and semi-colon, thus: κανών·
Question-mark = English semi-colon. τί; 'what?'
No mark of exclamation was in use.

SCRIPT

The method of writing Greek letters is as follows:

α β γ δ ε ζ η θ ι κ λ μ ν ξ ο π ρ σ (ς) τ υ φ χ ψ ω

The letters should be written as close together as possible but must not be joined. A space equal to one letter at least should be left between words.

Note which letters project above or below the normal.

Capital letters need not be used except for proper names.

Ex. 1. Copy out and translate, using the Index at the end of the book: ἐγώ εἰμι τὸ Ἄλφα καὶ τὸ Ὦ(μέγα), τὸ πρῶτον καὶ τὸ ἔσχατον.

EXERCISES IN WRITING AND TRANSLITERATION

The words are actual Greek words but in some instances the meaning has changed slightly in the course of 2000 years.

Remember that *κ* = c, *ῥ* = rh, *υ* = y,
χ = ch, *γ* before *γ*, *κ*, *χ*, *ξ* = n,
ει = i, *ου* = u, *οι* = oe or ē, *αι* = ae or ē.

Ex. 2. Write as English words, observing the long and short *e* and *o*:

κρισις	*Δελτα*	*Γενεσις*
πυθων	*πατηρ*	*κινημα*
Σοφια	*αὐτοματον*	*ὁριζων*
διλημμα	*χαρακτηρ*	*φαλαγξ*
ῥινοκερως	*Ξερξης*	*Κυκλωψ*

Ex. 3. Write as Greek words, observing the long and short marks:

basis	drama	climax
nĕctar	ēchō	astēr
Philadĕlphia	hērōĕs	ŏrchēstra
zōnē	isŏscĕlēs	Hĕllas
phthisis	lynx	panthēr

The Greek form will be found in the Index.

Ex. 4. What English words come from:

λιτανεια	*δημαγωγος*	*στρατηγος*
συμφωνος	*μουσικη*	*συγκοπη*
Σειρην	*γεωργος*	*διφθερα*
ἱπποποταμος	*γυμνασιον*	*σφαιρα*
Αἰγυπτος	*Φοινιξ*	*οἰκονομια*
ΣΦΙΓΞ	ΨΥΧΗ	ΛΗΘΗ

Afterwards correct these from the Index.

THE VERB

Greek verbs have a Dual number as well as a Singular and a Plural; a Middle voice as well as Active and Passive; an extra tense—the Aorist.

These differences will be dealt with later. At present the verb will be presented as in Latin. There are two conjugations: 1st person singular in -*ω* and in -*μι*.

VERBS IN -ω

Active Voice. Present Indicative of λύ-ω 'I loose'

Sing.	1. λύ-ω	I loose	*Plur.*	λύ-ομεν	we loose
	2. λύ-εις	you loose		λύ-ετε	you loose
	3. λύ-ει	he looses		λύ-ουσι(ν)[1]	they loose

The other meanings are 'I do loose'; 'I am loosing.'

Rule I. Never use the 2nd person plural for a single person. In Latin and Greek 'you,' 'yours,' were never used as in French and English to refer to a single person. Moreover the 2nd sing. is often irregularly formed and if not used will never be mastered.

Future tense is from the stem λυσ- with the same personal endings as the present: λύσομεν 'we shall' or 'will loose.'

Imperfect tense. All past (historic) tenses are marked by an ἐ- prefixed to the stem in those verbs which begin with a consonant. This is called the Syllabic Augment. *N.B.* ρ is doubled as ῥίπτω 'I hurl,' *Imperf.* ἔρριπτον.

Imperfect Indicative

ἔ-λυ-ον	I was loosing	ἐ-λύ-ομεν	we were loosing
ἔ-λυ-ες	you were loosing	ἐ-λύ-ετε	you were loosing
ἔ-λυ-ε(ν)[1]	he was loosing	ἔ-λυ-ον	they were loosing

The meaning is also 'I used to loose.'

Rule II. The Imperfect is used in many sentences denoting habitual, attempted or continuous action in past time, where English is content with the ordinary Past; especially for duration of time: *e.g.* They marched (ἐ-στράτευ-ον) all night.

The negative corresponding to 'non' in Latin is οὐ before a consonant, οὐ λύω 'I do not loose'; οὐχ before an aspirate, οὐχ ἱππεύω 'I do not ride'; οὐκ before an unaspirated vowel, οὐκ ἐθέλω 'I am not willing.'

Ex. 5. Write out with meanings the future and imperfect of πιστεύ-ω.

[1] The (ν) in brackets is added when the following word begins with a vowel or before a stop. The Greeks disliked what is called *hiatus*; *i.e.* a short vowel at the end of a word coming before a word beginning with a vowel. Other devices to avoid this were to cut off α ε ι ο as in the French 'qu', l'' for 'que, la'; or to arrange the words in a different order or by crasis (see p. 26).

ἱππεύ-ω ride.
χορεύ-ω dance.
φεύγ-ω flee, avoid.
πιστεύ-ω trust, believe (*dat.*).
οὐ, οὐκ, οὐχ not.
οὐδέν nothing, in no way (*adv.*).
πῶς; how?
*γάρ[1] for.
κελεύ-ω order, command.
πείθ-ω advise, persuade.
κωλύ-ω hinder, prevent.
ἐθέλ-ω wish, am willing.
καί also, even, both, and.
τί; what? why? (*adv.*).
ἀλλά but.
*οὖν[1] therefore.

Note. (1) The verbs in the right hand column are followed by an infinitive when required. If a negative is needed after the first two it will be μή.

(2) Present and future infinitive end in -ειν: λύειν, λύσειν.

Ex. 6. 1. φεύγομεν. τί φεύγετε; οὐκ ἔφευγεν.
2. ἐκέλευέ **με**[2] μὴ φεύγειν.
3. οὐκ ἐθέλει χορεύειν.
4. φεύγουσιν, οὐ γὰρ πιστεύουσί **μοι**[2].
5. ἔπειθόν με ἱππεύειν ἀλλ' οὐκ ἐθέλω.
6. πῶς κωλύσετέ με φεύγειν; οὐδὲν γάρ με κωλύσει.
7. πιστεύει μοι· φεύγειν οὖν οὐκ ἐθέλει.
8. οὐ κωλύουσί με φεύγειν, ἀλλὰ καὶ κελεύουσιν.

Ex. 7. 1. Why were they fleeing? They used to trust me.
2. We do not ride but we are willing to dance.
3. He will not trust me; therefore he flees.
4. You (*Rule I, p.* 4) order me to flee but I am unwilling.
5. Nothing therefore hinders me from-riding (*infin.*).
6. They do not trust me but even order me to flee.
7. He therefore will loose me and will order me not to flee.
8. Why do you not ride?

Always assume 'you' to be singular unless it **must** be plural.

[1] Particles marked thus stand second in the sentence as 'enim,' 'autem,' etc. in Latin.
[2] Words in bolder type should be guessed but will be found in the Index.

FIRST DECLENSION. FEMININE NOUNS -α, -η

Nouns have singular (*dual*, p. 88) and plural.

Feminine nouns end in -α or -η.

Rule III. Nouns in -α are of two types: (1) Those that end in -εα, -ια, -ρα retain α throughout the singular. This is called α pure and applies throughout Attic Greek—adjectives, numerals and verbs. (2) Other nouns in -α have genitive in -ης, dative in -ῃ. Nouns in -η retain η throughout the singular.

Sing.	the	land	sea	rule, beginning
Nom.	ἡ	χώρ-α	θάλαττ-α	ἀρχ-ή
Voc.	—	χώρ-α	θάλαττ-α	ἀρχ-ή
Acc.	τὴν	χώρ-αν	θάλαττ-αν	ἀρχ-ήν
Gen.	τῆς	χώρ-ας	θαλάττ-ης	ἀρχ-ῆς
Dat.	τῇ	χώρ-ᾳ	θαλάττ-ῃ	ἀρχ-ῇ

Plural. All first declension words, masc. and fem., have the same endings for the plural. The dative case always has an ι.

Nom.	αἱ	χῶρ-αι
Voc.	—	χῶρ-αι
Acc.	τὰς	χώρ-ας
Gen.	τῶν	χωρ-ῶν (for χωράων)
Dat.	ταῖς	χώρ-αις

(1) There is no ablative. The cases are used, generally, as in Latin but the uses of the ablative are divided between the genitive and the dative (with the help of prepositions). The dative especially expresses instrument, cause, means, time when, manner, as well as the Latin uses.

(2) When the subject of a verb is a pronoun it is usually omitted unless emphatic.

(3) All the commoner verbs have a variety of similar meanings, *e.g.*

λύει τὴν ἐπιστολήν	'he opens the letter'
ἔλυον τὴν γέφυραν	'they were breaking up the bridge'

The definite article is used in the same ways as in French; at present use it where it appears in English.

ἡ οἰκία house.
ἡ θύρα door.
ἡ ἀγορά market-place.
ἡ γῆ earth, land.
ἔχω have, hold.
κλείω shut.
οὔτε...οὔτε neither...nor.
ἐστί(ν) is.

ἡ μάχη battle.
ἡ κώμη village.
ἡ τράπεζα table.
εἰς (often ἐς) into (*acc.*).
ἐν in, on, among (*dat.*).
ἐκ from, out of (*gen.*).
ἐξ (before a vowel).
εἰσί(ν) they are.

N.B. ἔστι and εἰσί at the beginning of a sentence = there is, there are.

Ex. 8. 1. ἔστι μάχη ἐν τῇ θαλάττῃ. ἡ οἰκία ἔχει δύο θύρας.
2. ἐφεύγετε εἰς τὰς οἰκίας. ἐφεύγομεν ἐκ τῆς οἰκίας.
3. οὐκ ἔστι τράπεζα ἐν τῇ οἰκίᾳ. ἱππεύσομεν οὖν εἰς τὴν ἀγοράν.
4. Ἱπποκλείδης ἐχόρευε καὶ ἐν τῇ τραπέζῃ.
5. ἔχομεν γὰρ τὴν ἀρχὴν καὶ τῆς θαλάττης καὶ τῆς γῆς.
6. ἔκλειον τὰς θύρας· ἔστι γὰρ μάχη ἐν τῇ κώμῃ.
7. ἔπειθέ με μὴ κλείειν τὴν θύραν· ἔχομεν γὰρ οὐδὲν ἐν τῇ οἰκίᾳ.
8. πῶς οὐκ ἔστιν ἀγορὰ ἐν ταῖς Ἀθηναῖς;

N.B. πῶς οὐ...; expresses surprise at someone's ignorance. It may be translated 'of course'—or 'certainly.'

Ex. 9. 1. There are two market-places in the village. 2. We will shut the door. 3. Why are you (*Rule I, p.* 4) riding in the market-place of the village? 4. There are battles both in the land[1] and on the sea. 5. He was persuading me to flee from the village into the house. 6. We are fleeing, for there is a shindy in the village. 7. Neither in Athens nor in the village have I a house. 8. Nothing shall prevent me (from) dancing (*infin.*) on the table. 9. What have you in the house? We have nothing. 10. They have the rule of the land but not of the sea. 11. I will shut the doors of the houses. What do you wish? 12. They were trusting me in the beginning. 13. They will ride even into the villages; therefore we are closing the doors of the houses.

Make a list of the mute consonants.

[1] Use γῆ for land when contrasted with sea or sky, and for earth; use χώρα otherwise.

INDICATIVE ACTIVE OF λύω

	Aorist	*Perfect*
Singular.	ἔ-λυ-σα I loosed, I did loose	λέ-λυ-κα I have loosed
	ἔ-λυ-σας	λέ-λυ-κας
	ἔ-λυ-σε(ν)	λέ-λυ-κε(ν)
Plural.	ἐ-λύ-σαμεν	λε-λύ-καμεν
	ἐ-λύ-σατε	λε-λύ-κατε
	ἔ-λυ-σαν	λε-λύ-κασι(ν)
Infinitive.	λῦσαι	λε-λυ-κέναι

The Aorist tense expresses an occurrence in past time.

The Perfect expresses an action as already finished at the present time and therefore is a primary tense.

The Aorist and the Perfect must be kept quite distinct.

ἔ-κλει-σα τὴν θύραν 'I closed the door, I did shut the door (but someone may have opened it since).'

κέ-κλει-κα τὴν θύραν 'I *have* shut the door (and it is still shut).'

A verb beginning with a single consonant or a mute followed by λ or ρ forms the perfect stem by repeating the first consonant with ε: πιστεύ-ω πεπίστευκα, κλεί-ω κέκλεικα.

FIRST DECLENSION. MASCULINE NOUNS IN -ας, -ης

	the	youth	judge
Nom.	ὁ	νεανί-ας	κριτ-ής
Voc.	—	νεανί-α	κριτ-ά
Acc.	τὸν	νεανί-αν	κριτ-ήν
Gen.	τοῦ	νεανί-ου	κριτ-οῦ
Dat.	τῷ	νεανί-ᾳ	κριτ-ῇ

Plural endings are the same as for feminine nouns.

Nom.	οἱ	νεανί-αι	κριταί
Voc.	—	νεανί-αι	κριταί
Acc.	τοὺς	νεανί-ας	κριτάς
Gen.	τῶν	νεανι-ῶν	κριτῶν
Dat.	τοῖς	νεανί-αις	κριταῖς

Patronymics—meaning 'son of' or any 'descendant of'—have vocative in -η:

e.g. Κρονίδης 'son of Κρόνος,' *Voc.* Κρονίδη.

Vocatives are usually preceded by ὦ.

ὁ κριτής judge.	παιδεύ-ω educate, train.
ὁ στρατιώτης soldier	πέμπ-ω send.
ὁ ναύτης sailor.	ἡ νίκη victory.
ὁ πολίτης citizen.	ἐπεί when, since.
ὁ ποιητής poet.	εὖ *adv.* well.
ὁ κλέπτης thief.	νῦν *adv.* now.

πρό *preposition* (*genitive case*), 'in front of, before' (of time).

(1) The article is used for the possessive adjectives ('my,' 'your,' 'their' etc.) where no confusion is likely to result.

(2) The negative and adverbs should be placed before the word that they affect: 'to train well' εὖ παιδεύειν.

(3) Names of whole classes, *e.g.* 'poets,' 'soldiers,' must have the article.

Ex. 10. 1. οὔτε οἱ ποιηταὶ οὔτε οἱ ναῦται εὖ ἱππεύουσιν.

2. Λυκοῦργος ὁ κριτὴς πεπαίδευκε τοὺς νεανίας τῆς κώμης.

3. ἐπεὶ οὖν τοῖς στρατιώταις οὐκ ἐπίστευσεν, ἔκλεισε τὴν θύραν.

4. ὦ κριτά, τί τοὺς κλέπτας οὐ πέμπεις ἐκ τῆς χώρας;

5. οἱ νεανίαι χορεύσουσιν ἐν τῇ ἀγορᾷ τῶν Ἀθηνῶν, ἔχομεν γὰρ νίκην ἐν τῇ ναυμαχίᾳ (sea-fight).

6. πρὸ τῆς ναυμαχίας ἐπαιδεύομεν τοὺς ναύτας.

7. ἐπεὶ οὐδὲν ἔχεις ἐν τῇ οἰκίᾳ, τί κέκλεικας τὴν θύραν;

8. ἐν τῇ οἰκίᾳ τοῦ ποιητοῦ ἔστι τράπεζα καὶ οὐδὲν ἄλλο.

9. ἐθέλομεν οὖν πέμπειν ἐπιστολήν, ἡ γὰρ νίκη λέλυκε τοὺς πολίτας.

Ex. 11. 1. O citizens, you have a victory both on the sea and on the land. 2. The judge has ordered the lads of the village not to dance in the market-place in front of his house. 3. Why did you not shut the door? There are now burglars in the house. 4. Since you (*Rule I, p.* 4) did not train the young man well, O Simōnides, he is now a thief. 5. We wish to have a victory, therefore we are sending neither poets nor judges but soldiers. 6. They were trying-to-persuade (*imperf.*) the judge not to set the thief free. 7. When you set free the young men before the sea-fight you prevented a victory.

Note: 'in the poet's house' can be arranged in three ways in Greek: (*a*) ἐν τῇ οἰκίᾳ τοῦ ποιητοῦ, (*b*) ἐν τῇ τοῦ ποιητοῦ οἰκίᾳ, (*c*) ἐν τῇ οἰκίᾳ τῇ τοῦ ποιητοῦ. (*b*) is the usual order for qualifying genitives, (*c*) is used when there is an adjective or if the phrase is a long one, (*a*) is the order for partitive expressions.

Pluperfect Indic. Act.	εἶναι to be *Pres. Indic.*	*Imperf. Indic.*
I had loosed	I am	I was
ἐ-λε-λύ-κη (-κειν)	εἰμί ('sum')	ἦν or ἦ
ἐ-λε-λύ-κης (-κεις)	εἶ	ἦσθα
ἐ-λε-λύ-κει	ἐσ-τί(ν)	ἦν
ἐ-λε-λύ-κεμεν (-κειμεν)	ἐσ-μέν	ἦμεν
ἐ-λε-λύ-κετε (-κειτε)	ἐσ-τέ	ἦτε
ἐ-λε-λύ-κεσαν (-κεισαν)	εἰσί(ν)	ἦσαν

INFINITIVE

Rule IV. The chief use of the infinitive is to represent reported statements and commands. The tense used must be the same as that of the original unreported sentence. In Latin the subject of the infinitive is always accusative but in Greek it is omitted if the speaker refers to himself; or if required, will be in the nc minative case, *e.g.*

He said that the soldiers were fleeing = ἔφη τοὺς στρατιώτας φεύγειν,
but he said that he wished to flee (himself) = ἔφη ἐθέλειν φεύγειν (αὐτός).

(οὐκ) ἔφη εἶναι κριτής = he said that he was (not) a judge.

Note that the negative is placed before ἔφη (cp. 'nego' for 'dico...non'). This does not apply to other verbs of saying. ἔφη is also the verb to use ('inquit') in the middle of a speaker's actual words: Ὦ πολῖται, ἔφη, ἔστιν ἡμῖν νίκη καὶ κατὰ γῆν καὶ κατὰ θάλατταν, '"We are victorious, citizens," said he, "both on land and sea."' As in Latin the verb 'to be' with the dative is often used for the possessor.

There is no infinitive form in English for the future or aorist tense because we do not change the mood but the tense: compare

κλείσω τὴν θύραν I *will* shut the door.	ἔφη κλείσειν τὴν θύραν He said he *would* shut the door.
ἔκλεισα τὴν θύραν I closed the door.	ἔφη κλεῖσαι τὴν θύραν He said that he closed the door.

ὁ τοξότης archer.	*οἱ τοξόται* the police.
ἔφη he said, said he.	*ἤδη* already, now, soon ('iam').
ἔφασαν they said, said they.	*κρού-ω* I knock, strike.
οὐ μόνον not only.	*ἀλλὰ καί* but also.

Rule V. *μὲν...*δέ are used with words and sentences, where the contrast is slight, *i.e.* 'on the one hand...on the other hand'; *μέν* need not be translated. *δέ* is often used alone meaning 'but,' 'and,' or merely 'now,' 'to resume,' etc. Sentences are always connected in some way; usually in English this is left to the sense.

Ex. 12. 1. *ἐν δὲ τῇ μάχῃ οἱ μὲν στρατιῶται ἔφευγον οἱ δὲ ναῦται ἔχουσι τὴν νίκην.* 2. *οἱ δὲ πολῖται ἤδη ἐλελύκεσαν τὸν κριτήν.* 3. *ἔφη εὖ πεπαιδευκέναι τοὺς νεανίας οὐ μόνον εἰς τὴν μουσικὴν ἀλλὰ καὶ εἰς τὴν μαθηματικήν.* 4. *οἱ τοξόται, ἔφη, ἤδη κρούουσι τὴν θύραν.* 5. *ἐπεὶ οὖν ἔκρουσα τὴν θύραν, ὁ μὲν ποιητὴς οὐκ ἔφη εἶναι ἐν τῇ οἰκίᾳ, ἐγὼ δ' οὐκ ἐπίστευσα τῷ ποιητῇ.* 6. *ὁ μὲν κριτὴς ἦν ἐν τῇ οἰκίᾳ· οἱ δὲ νεανίαι ἔκρουον τὴν θύραν καὶ ἔφευγον.* 7. *ἔπεμψαν τοὺς τοξότας· ἔφασαν γὰρ εἶναι κλέπτας ἐν τῇ οἰκίᾳ τῇ τοῦ πολίτου.* 8. *οὐκ ἔφη εἶναι ποιητής, ἀλλὰ κριτὴς τῶν ποιητῶν.*

Ex. 13. 1. The judge therefore was sending the soldiers into the market-place. 2. He had already freed not only the young man but also the poet. 3. You have not trained the young man well, o judge, for he is now striking the poet. 4. The soldier (*Rule V, above*) denied (that he) was a burglar but the judges trusted the police since he was in the citizen's house. 5. The citizens were not fleeing; for, said they, we trust the sailors. 6. Sailors (*p.* 9 (3)) have command of the sea, soldiers of the land.

ORDER

The order of a sentence is the most natural arrangement much as in English but a word in an unusual position receives emphasis. The position of the article is the most important point. The verb should not be placed at the end of a clause unless it comes there naturally.

NEGATIVES

Negatives do not cancel each other unless a compound negative precedes *οὐ*, *e.g.* *οὐ πέμπει οὐδέν* 'he sends nothing': but *οὐδὲν οὐ πέμπει* would mean 'he sends everything.'

For emphasis several negatives are often used in one sentence.

SECOND DECLENSION. MASCULINE AND FEMININE NOUNS IN -ος, NEUTER IN -ον

speech, word		gift	
ὁ λόγ-ος	οἱ λόγ-οι	τὸ δῶρ-ον	τὰ δῶρ-α
— λόγ-ε	— λόγ-οι	— δῶρ-ον	— δῶρ-α
τὸν λόγ-ον	τοὺς λόγ-ους	τὸ δῶρ-ον	τὰ δῶρ-α
τοῦ λόγ-ου	τῶν λόγ-ων	τοῦ δώρ-ου	τῶν δώρ-ων
τῷ λόγ-ῳ	τοῖς λόγ-οις	τῷ δώρ-ῳ	τοῖς δώρ-οις

As in Latin, neuter words have plural in -α and *Nom. Voc. Acc.* the same but (*Rule VI*) a neuter plural subject has a verb in the 3rd person singular.

There are a number of feminine nouns declined like λόγος.

TEMPORAL AUGMENT

When the verb begins with a vowel, the vowel is lengthened in the Historic tenses of the Indicative:

α and ε	become	η	ἄγω	lead,	*imperf.*	ἦγον
			ἐλαύνω	drive	„	ἤλαυνον
ο	becomes	ω	ὁπλίζω	arm	„	ὥπλιζον
ι	„	ῑ	ἱππεύω	ride	„	ἵππευον
υ	„	ῡ	ὑβρίζω	insult	„	ὕβριζον
αι	„	ῃ	αἴρω	lift	„	ᾖρον
οι	„	ῳ	οἰκτείρω	pity	„	ᾤκτειρον

REDUPLICATION

The same lengthening is used instead of reduplication for the perfect of these verbs but is retained in all the moods, infinitive and participle.

PREPOSITIONS

Some prepositions are used with two and some with three cases with difference in meaning. These will be learnt by degrees: *e.g.* παρά:

ἡ οἰκία ἐστὶ παρὰ τῇ θαλάττῃ
The house is beside the sea.
ἐπέμπομεν στρατιώτας παρὰ τὸν κριτήν
We were sending soldiers to the judge.
τὰ δῶρα ἦν παρὰ τοῦ κριτοῦ
The gifts were from the judge.

With two cases

μετά with acc. = after; with gen. = with (accompanied by).
διά with acc. = on account of; with gen. = through, across.

ὁ νόμος law, custom.	ἡ νῆσος island.
ὁ στρατηγός general.	τὸ ὅπλον (*pl.*) arms.
ὁ στρατός army.	τὸ πλοῖον ship.
ὁ δοῦλος slave.	τὸ πεδίον plain.
ὁ δῆμος people.	τὸ χωρίον place, spot.
ὁ πόλεμος war.	στρατεύ-ω I march.
οἱ πολέμιοι the enemy.	ὁπλίζ-ω I arm.
ἡ ὁδός road, way.	ἐπί against (*acc.*).

Note. The article ὁ ἡ τό is often used (with noun understood) before an adjective, an adverb, or a preposition-phrase, *e.g.* οἱ ἐν τῇ νήσῳ 'the men on the island.' τὰ τοῦ κριτοῦ 'the affairs of the judge.' οἱ νῦν 'the men of the present day.' ὁ or ἡ with the genitive of a proper name = 'the son of, the daughter of': Κῦρος ὁ Δαρείου 'Cyrus son of Darius.'

Ex. 14. 1. ὁ μὲν στρατηγὸς ἦγε τὸν στρατὸν διὰ τοῦ πεδίου, οἱ δὲ πολέμιοι ἔπεμπον τὰ πλοῖα καὶ τοὺς ναύτας ἐπὶ τὰς νήσους.

2. μετὰ τὸν πόλεμον οἱ πολῖται ἔπειθον δώροις τὸν κριτήν· ὁ γὰρ δῆμος νῦν οὐ πιστεύει τοῖς νόμοις.
3. τὰ δῶρά ἐστιν ἐν τῇ οἰκίᾳ· τί οὖν οὐκ ἔκλεισας τὴν θύραν;
4. ἐπεὶ οἱ πολέμιοι ἐστράτευον εἰς τὸ χωρίον, ὡπλίζομεν τοὺς νεανίας· ᾠκτείρομεν γὰρ τοὺς ἐν τῇ κώμῃ.
5. οἱ μετὰ τοῦ στρατηγοῦ ἦσαν ἐν τῇ ὁδῷ πρὸ τῆς τοῦ κριτοῦ οἰκίας.
6. νόμος ἦν τοῖς πολίταις μετὰ τὴν νίκην χορεύειν ἐν τῇ ἀγορᾷ.
7. μετὰ δὲ τὴν μάχην Μιλτιάδης ὁ Κυψέλου ἦν ἐν ταῖς Ἀθήναις μετὰ τοῦ στρατοῦ.

Ex. 15. 1. Before the war (*Rule V, p.* 11) the laws used to prevent the young men from dancing in the road but now even the police wish to dance.

2. They said that on account of the war the judge had already loosed both the slaves and the robbers.
3. In the war the soldiers were marching into the villages while those in the islands were arming the sailors.
4. Since the arms are (*Rule VI, p.* 12) already on (board of) the ships we have ordered the police to arm the citizens.
5. The judge by his words was persuading the people (*singular*) to trust the laws.
6. Parse ὡπλίζομεν, λελυκέναι, λῦσαι, ἦτε, ἐστρατεύκει.

ADJECTIVES IN -ος

As in Latin, the masc. and neuter follow the 2nd decl., the fem. follows the 1st decl. but *Rule III* applies, *i.e.* adjectives in -εος, -ιος, -ρος will have α throughout the singular like χώρα.

Rule VII. Compound adjectives (formed from two words) have fem. the same as the masc. *e.g.* ἄνομος -ον 'lawless,' εὔλογος -ον 'reasonable.'

Adverbs of manner answering the question πῶς; 'how?' are formed from adjectives by changing the last syllable of the gen. sing. masc. into -ως, *e.g.* ἀνόμως 'lawlessly'; καλῶς 'nobly, well.'

καλός 'beautiful, honourable, fine.'

Sing.	*Nom.*	καλ-ός	-ή	-όν	*Plur.* καλ-οί	-αί	-ά
	Voc.	καλ-έ	-ή	-όν	καλ-οί	-αί	-ά
	Acc.	καλ-όν	-ήν	-όν	καλ-ούς	-άς	-ά
	Gen.	καλ-οῦ	-ῆς	-οῦ	καλ-ῶν	-ῶν	-ῶν
	Dat.	καλ-ῷ	-ῇ	-ῷ	καλ-οῖς	-αῖς	-οῖς

αἰσχρός 'ugly, disgraceful.'

Fem. by Rule III (*p.* 6): αἰσχρά, αἰσχράν, αἰσχρᾶς, αἰσχρᾷ.

The adjectives πολύς 'much' (pl. 'many'), μέγας 'great,' are irregular in the sing. Plural regular from stems πολλ-, μεγαλ-.

Nom.	πολύς	πολλή	πολύ	μέγας	μεγάλη	μέγα
Acc.	πολύν	πολλήν	πολύ	μέγαν	μεγάλην	μέγα
Gen.	πολλοῦ	πολλῆς	πολλοῦ	μεγάλου	μεγάλης	μεγάλου
Dat.	πολλῷ	πολλῇ	πολλῷ	μεγάλῳ	μεγάλῃ	μεγάλῳ

Note. (1) The position of the adj. is generally speaking the same as in English: ἡ καλὴ νῆσος 'the beautiful isle,' ἡ νῆσός ἐστι καλή (or καλὴ ἡ νῆσος) 'the island is beautiful.'

(2) An adjective is converted into a noun by prefixing the article: *e.g.* τὸ καλόν 'beauty.' οἱ πολλοί 'the crowd.' τὰ αἰσχρά 'all that is vile.'

(3) *Note* οἱ μέν 'some,' οἱ δέ 'others,' declined throughout sing. and plural.

(4) The adverb of πολύς is πολύ, the neuter, and so with other adjectives of quantity such as ὀλίγος 'little.' μάλα is generally used for μεγάλως.

(5) Adjectives of position such as μέσος 'middle,' ἄκρος 'top,' are often placed first: ἐν μέσῃ τῇ νήσῳ 'in the middle of the island.'

WORD-FORMATION

The English language, especially the scientific part of it, has borrowed many Greek words: (1) to avoid using familiar names already fixed in meaning, (2) because compound words are easily formed in Greek.

The great majority are formed by joining two or more stems with the connecting vowel *o* and by giving the endings *-ic* or *-ical* for an adjective, *-ia* or *-y* for an abstract noun, a mute *e* for an ordinary noun: 'demo-cratic' from δῆμος 'people,' and κράτος 'power'; 'autobiography' from αὐτός 'self,' βίος 'life,' γράφω 'I write'; 'hemisphere' from ἡμι- 'half,' σφαῖρα 'a ball.' If the first word already ends in a short vowel, it is retained: 'telephone' τῆλε 'far off,' φωνή 'sound'; 'Polynesia' πολύς 'many,' νῆσος 'island.'

An extensive Greek vocabulary is soon acquired, as in most cases a single stem produces all the words connected with one notion. Contrast with English and Latin:

στρατός	army	exercitus
στρατηγός	general	imperator
στρατήγιον	headquarters	praetorium
στρατιώτης	soldier	miles
στρατόπεδον	camp	castra
στρατοπεδεύω	encamp	castra pono
στρατεύω	march to war	bellum infero
etc.		

So all the words connected with a law-court from the stem δικ-:

δίκη	justice
δικάζω	give judgment
δικαστής	juryman
δικαστήριον	law-court
δίκαιος	just
ἄδικος	unjust
ἀδικέω	injure

Verbs are freely compounded with prepositions: ἀπο-πέμπειν 'to send away,' δι-εκ-πλέω 'I sail out through.' σύν 'with, together,' συ-στρατεύ-ω 'to march with,' συγ-κλείω 'shut up.'

TIME AND SPACE

Duration of time, *acc.*; *πολὺν χρόνον* 'for a long time'; but with a negative, *gen.* would be used: *εἰκόσιν ἐτῶν οὐκ ἀπεδήμησα* 'for twenty years I never was away.'

Definite time when, *dat.*: *τῇ τρίτῃ ἡμέρᾳ* 'on the third day.' Time within which, *gen.*: *ἡμέρας* 'by day,' *τριῶν ἡμερῶν* 'within three days.'

A preposition is sometimes used to make things clearer:

ἐπὶ τοῦ Κύρου	in the days of (in the reign of) Cyrus.
δι' ὀλίγου (χρόνου)	after a short interval.
ὑπὸ τὸν σεισμόν	about the time of the earthquake.

Extent of space, *acc.*; *ἀπέχει τῶν Θηβῶν ἡ Πλάταια σταδίους ἑβδομήκοντα* 'Plataea is 70 furlongs distant from Thebes.'

ὁ ποταμὸς τρία ἔχει στάδια τὸ εὖρος 'the river is three furlongs in breadth.' *στάδιον* in the plural is either masc. or neuter.

IMPERSONAL VERBS

δεῖ, imperf. ἔδει, infin. δεῖν, 'it is necessary' = 'must.'

χρή, imperf. ἐχρῆν, infin. χρῆναι, 'ought.'

These two have acc. and infin. ('oportet' in Latin).

ἔξεστι, imperf. ἐξῆν, infin. ἐξεῖναι, 'it is allowed'; *with dat.*

So also a neuter adjective and the verb 'to be': *ἀδύνατόν ἐστι* 'it is impossible.'

Questions are often asked by *ἆρα* (Latin '-ne') merely inquiring.

ἆρ' οὐ and *οὔκουν* ('nonne'). Answer expected 'yes' = 'surely.'

ἆρα μὴ ('num'). Answer expected 'no' = 'surely not.'

Ex. 16. 1. ἆρ' οὐ δεῖ τοὺς στρατιώτας πιστεύειν τῷ στρατηγῷ;
2. ἐκέλευσεν οὖν τὸν δοῦλον πέμπειν τὴν ἐπιστολὴν δι' ὀλίγου.
3. τῇ δὲ δευτέρᾳ ἡμέρᾳ ἔπεμψαν πολλὰ καὶ καλὰ πλοῖα.
4. ἆρ' ἀδύνατόν ἐστι πείθειν τοὺς πολίτας λύειν τὸν κριτήν;
5. πολλὰς οὖν ἡμέρας ἐξῆν τῷ δήμῳ εἶναι ἐν τῷ δικαστηρίῳ.

Ex. 17. 1. Within three days therefore we will set free the men in the islands. 2. Ought not the general to arm the men in the villages? 3. He said that the house was two furlongs distant from the great sea. 4. In the days of Miltiades there were many *and* great battles. 5. On the third day they encamped beside the great river.

δίκαι-ος -α -ον just.
ἄξι-ος -α -ον worthy.
δεύτερος -α -ον second.
τρίτ-ος -η -ον third.
πρῶτος -η -ον first.
πρῶτον *adv.* first.
κακ-ός -ή -όν bad, cowardly.
σοφ-ός -ή -όν wise.
πιστ-ός -ή -όν faithful.
ὀλίγ-ος -η -ον little, *pl.* few.
χρήσιμ-ος -η -ον useful.
ὁ ἵππος horse (*fem.* cavalry).
ἀπό *prep. with gen.* from.
ἔπειτα *adv.* then, afterwards.
ἀγαθ-ός -ή -όν good, brave.
ὁ χρόνος time.

Ex. 18. 1. οἱ πιστοὶ δοῦλοί εἰσιν ἄξιοι πολλῶν καὶ καλῶν δώρων.

2. πολὺν μὲν χρόνον ἔφευγον οἱ πολέμιοι ἀπὸ τῆς κώμης· νῦν δ' εἰσὶν ἐν τῷ μεγάλῳ πεδίῳ.

3. τί νῦν οὐκ ἀποφεύγετε; ἔχετε γὰρ καλοὺς ἵππους.

4. δύο μὲν ἡμέρας ἦσαν οἱ στρατιῶται ἐν τῇ τοῦ ποιητοῦ οἰκίᾳ, τῇ δὲ τρίτῃ ἔφευγον ἐκ τῆς κώμης.

5. οἱ καλοὶ λόγοι οἱ τοῦ ἀγαθοῦ κριτοῦ ἔπειθον τοὺς πολίτας.

6. τὰ πλοῖα ἦν χρήσιμα τοῖς στρατιώταις· οἱ γὰρ ἐν ταῖς νήσοις ἀνόμως ὕβριζον τὸν στρατηγόν.

7. οἱ ἀπὸ Κορίνθου πρῶτον μὲν ἵππευσαν εἰς τὴν ἀγοράν, ἔπειτα δὲ μετὰ τῶν ἀνόμων πολιτῶν ἤλαυνον τοὺς ἵππους.

8. πρῶτος ἐγὼ (was the first to) ἐστράτευσα ἐπὶ τὴν τῶν πολεμίων χώραν.

9. δίκαιος καὶ σόφος ὁ στρατηγός· ἔφη οὖν λύσειν τοὺς πολίτας.

10. πολλοὶ μὲν ἐθέλουσι χορεύειν· ὀλίγοι δ' εὖ χορεύουσιν.

11. τί κλείεις νῦν τὴν τοῦ σταθμοῦ θύραν ἐπεὶ ὁ κλέπτης ἤδη ἄγει τὸν ἵππον διὰ τοῦ πεδίου;

Ex. 19. 1. After the great war for a long time there were many robbers in the land. 2. It is disgraceful to insult the poet; surely (*p.* 7, *N.B.*) he is useful to Athens. 3. The village is lawless (*Rule VII, p.* 14); therefore we are sending many *tall* policemen. 4. For two days they rode (*Rule II, p.* 4) across the great plain but on the third on account of the battle they marched into a faithful village. 5. O cowardly slave, why did you not prevent the soldier from driving away the horses? You are not worthy even of a *small* gift. 6. For many days after the battle we were in a good place but afterwards we rode from the village into the plain beside the sea. 7. Ought not the army to trust a good general?

COMPARISON

Most adjectives form Comparative in -*τερος*, Superlative in -*τατος*.

When the adj. ends in -*ος* the *ο* will be lengthened to *ω* if the preceding syllable is short:

wise	*σόφος*	*σοφώτερος*	*σοφώτατος*
desolate	*ἔρημος*	*ἐρημότερος*	*ἐρημότατος*

Diphthongs, long vowels, and two consonants (the second not being *λ* or *ρ*) constitute a long syllable.

Adverbs in -*ως* take the neuter sing. of the Comparative and the neuter plural of the Superlative for their degrees:

ἀνόμως 'lawlessly,' *ἀνομώτερον*, *ἀνομώτατα*.

Sing. N. V.	*ἐγώ* 'I'	*Pl.* *ἡμεῖς*	*Sing.* *σύ* 'thou'	*Pl.* *ὑμεῖς*
Acc.	*ἐμέ*, *με*	*ἡμᾶς*	*σέ*	*ὑμᾶς*
Gen.	*ἐμοῦ*, *μου*	*ἡμῶν*	*σοῦ*	*ὑμῶν*
Dat.	*ἐμοί*, *μοι*	*ἡμῖν*	*σοί*	*ὑμῖν*

(1) The forms **με*, **μου*, **μοι* must not begin a clause or be used after a preposition. The vowel of the preposition is elided (except of *περί* and *πρό*) thus: 'with me' *μετ' ἐμοῦ*; 'after you' (*pl.*) *μεθ' ὑμᾶς*. *τ* and *π* are aspirated before an aspirate. *ἐφ' ἵππου* 'on horseback.'

(2) After a comparative, nom. or acc., 'than' will be translated by *ἤ*, but if the nouns are directly compared, the genitive can be used (where abl. in Latin). *ἤ*, however, cannot be omitted before a genitive or a dative.

(3) The relative pronoun *ὅς*, *ἥ*, *ὅ* (all pronouns drop the *ν* in the neuter) is declined like *καλός* and agrees as in Latin. The suffix -*περ* is sometimes added for emphasis, *ὅπερ* 'just what.'

(4) The *amount by which* one thing is greater or less is put in the dative. 'Much wiser' = *πολλῷ σοφώτερος*.

(5) *ὁ δέ* (not preceded by *ὁ μέν*) is used to refer to someone just mentioned who is *not* the subject of the last sentence: *ἐκέλευσε τὸν πολίτην πέμπειν τὴν ἐπιστολήν· ὁ δὲ οὐκ ἤθελε* 'He ordered the citizen to send the letter but he would not.'

If the subject of both clauses is the same, then *ἀλλά* will be used: *οἱ πολῖται ἐθέλουσι στρατεύειν ἀλλ' οὐκ ἔχουσιν ὅπλα* 'The citizens are willing to march but they have no arms.'

πλούσ-ι-ος -α -ον rich.
ἰσχῡρ-ός -ά -όν strong.
δειν-ός -ή -όν terrible, (*with infin.*) clever at.
ὁ ποταμός river.
*ποτέ once (upon a time).
ἀνά *prep. with acc.* up
χαλεπ-ός -ή -όν difficult, harsh.
ἄνευ *prep. with gen.* without.
πολλάκις *adv.* often.
πρός *prep. with acc.* towards, with a view to; *with dat.* near, in addition to; *with gen.* from the side of.
ὅς, ἥ, ὅ *relative pronoun* who.

Ex. 20. 1. ἐν τῷ πολέμῳ τὰ πλοῖά ἐστι χρησιμώτερα τοῖς ναύταις ἢ τῷ στρατηγῷ.
2. ὁ νεανίας ἔφη εἶναι πολλῷ σοφώτερος τοῦ κριτοῦ.
3. οἰκτείρομεν σὲ πολύ, ὦ πολῖτα, ὃς οὐκ ἔχεις οὔτε ὅπλα οὔτε ἵππον· ἡ γὰρ μάχη ἐν τῇ κώμῃ ἐστὶ δεινοτάτη.
4. ὁ κακὸς ἀθλητὴς ἔκρουσε τὸν ποιητήν· ὁ δὲ οὐκ ἤθελε φεύγειν.
5. ὁ μὲν δῆμος ἐκέλευσε τοὺς πλουσιωτάτους τῶν πολιτῶν πέμπειν ἄξια δῶρα πρὸς τὸν δίκαιον κριτήν· οἱ δ' οὐκ ἤθελον.
6. "ἐπεὶ ἵππευον ἀνὰ τὸ τοῦ Νείλου Δέλτα," ἔφη ὁ ὁδίτης, "ἦσαν ἐν τῷ ποταμῷ μεγάλοι κροκόδειλοι οἳ πρῶτον μὲν ᾖρον τὰς δεινὰς οὐρὰς ἐκ τοῦ ποταμοῦ, ἔπειτα δὲ ἔκρουον τὴν γῆν· ἐγὼ δ' ἔφευγον· πῶς γὰρ οὔ; τί οὖν οὐ πιστεύετέ μοι; εἰμὶ γὰρ ἀξιο-πιστότατος."

Ex. 21. 1. Soldiers (*p.* 9 (3)) are more useful in war to us than sailors. 2. The judge is very just; therefore he will prevent the rich citizen from driving the worthy young-man out of the land. 3. Many of the strongest youths were most justly driving the cowardly thieves across the plain. 4. After the victory it was very difficult to persuade the citizens to pity the enemy; for they said that the war was unjust. 5. *You* are richer than I, but I have many faithful slaves who will prevent you from striking me. 6. On the second day the battle was more terrible than on the first. 7. The general was leading us towards a strong place in which the enemy *formerly* were.

INDIRECT STATEMENT

φημί 'I say,' and οἶμαι (οἴομαι) 'I think,' take Infinitive, but other verbs of saying and thinking are often followed by ὅτι or ὡς (= English 'that') with the Indicative. The tense must not be changed as is done in English:

λέγουσιν ὅτι ἡ χώρα ἐστὶν ἐρημοτάτη
They say that the land is quite desolate.

ἔλεγον ὅτι ἡ χώρα ἐστὶν ἐρημοτάτη
They said that the land was quite desolate.

DEMONSTRATIVES

Nom. ὅδε, ἥδε, τόδε 'this (near me)' = the article + δε.
Acc. τόνδε, τήνδε, τόδε κ.τ.λ. (καὶ τὰ λοιπά = etcetera).
Nom. ἐκεῖνος, ἐκείνη, ἐκεῖνο κ.τ.λ. like καλός 'that (yonder).'

Sing.	οὗτος	αὕτη	τοῦτο	*Plur.*	οὗτοι	αὗται	ταῦτα
	τοῦτον	ταύτην	τοῦτο		τούτους	ταύτας	ταῦτα
	τούτου	ταύτης	τούτου		τούτων	τούτων	τούτων
	τούτῳ	ταύτῃ	τούτῳ		τούτοις	ταύταις	τούτοις

Meaning 'this or that (near you).'

Note: When there is an *ο* or *ω* in the final syllable there will be an *ο* in the stem; otherwise an *α*. The endings are the same as ὁ, ἡ, τό.

These words are used

(*a*) without a noun = he, she, it;

(*b*) in apposition to a noun, which *must* have the article:

on that day ταύτῃ τῇ ἡμέρᾳ or τῇ ἡμέρᾳ ταύτῃ.
this village ἥδε ἡ κώμη or ἡ κώμη ἥδε.
those arms ἐκεῖνα τὰ ὅπλα or τὰ ὅπλα ἐκεῖνα.

οὗτος (and its compounds) refer back; ὅδε applies to what follows:

ταῦτα μὲν ἔπειθε· ἐποίησε δὲ τάδε
That was his advice but he did as follows.

So too the adverbs οὕτω(ς) and ὧδε.

Special phrases: διὰ ταῦτα on that account; ἐν τούτῳ in the meanwhile; μετὰ ταῦτα after that; πρὸς τούτοις in addition to this; ἐπὶ τοῖσδε on the following terms.

THE RELATIVE

(1) The antecedent, if a demonstrative, is often omitted and the relative is governed by the preposition; ἀνθ' ὧν = ἀντὶ τούτων ἅ = in return for what; περὶ ὧν = about what; ἐξ οὗ = since the time when; διό = διὰ ὅ = on account of which thing, wherefore.

(2) When the antecedent is genitive or dative, the relative often is attracted into that case: πιστεύει οἷς ἔχει ὅπλοις 'he trusts what arms he has.'

ἄλλ-ος -η -ο other, else.

κατά *prep. with gen.* down from; *with acc.* according to (a person), along (a thing).

λέγω I say.

νομίζω reckon, think.

αἱ σπονδαί truce, treaty.

ἀκούω hear, listen to (*acc. of sound, gen. of cause*).

οὕτω(ς), ὧδε thus, so.

ἄρχω I rule (*gen.*).

βάρβαρος -ον foreign.

N.B. ἄλλος = Latin 'alius.' ἄλλοι...ἄλλοι 'some...others,' *but* ἄλλοι alone cannot mean 'some'; that would be ἔνιοι. οἱ ἄλλοι 'the others, the rest.' ἄλλοι ἄλλα ἔλεγον 'some said one thing some said another.' τἆλλα = 'in other respects.' ἄλλως = 'otherwise.'

Ex. 22. 1. ἐπεὶ ταῦτα ἤκουσεν, ὁ στρατηγὸς ἔφη τοὺς βαρβάρους λελυκέναι τὰς σπονδάς.

2. τῶν πολιτῶν οἱ μὲν ἤκουον τοῦ κριτοῦ, οἱ δὲ ἐν τούτῳ ὥπλιζον τοὺς νεανίας πρὸς τὴν ἐν τῇ ὁδῷ μάχην.

3. οὗτοι μὲν οἱ πολῖταί εἰσι χρήσιμοι ἡμῖν, ἐκείνους δὲ νομίζομεν εἶναι αἰσχροὺς κλέπτας· σὺ δὲ τί λέγεις;

4. λέγουσιν ὅτι ἐν ταύτῃ τῇ μάχῃ τῶν στρατιωτῶν ἄλλοι ἔφευγον πρὸς τὸ στρατόπεδον, ἄλλοι ἐνόμιζον εἶναι αἰσχρὸν φεύγειν.

5. ἄλλος ἄλλο λέγει· ἐγὼ δὲ οὐ πιστεύω τοῖς τῶν πολιτῶν λόγοις.

6. κατὰ τὸν Ἡρόδοτον νόμος ἦν ποτὲ ἐνίοις τῶν Αἰγυπτίων ἔχειν μικροὺς κροκοδείλους ἐν ταῖς οἰκίαις.

7. ἐπεὶ οὖν ἀδύνατόν ἐστιν ἡμῖν πέμπειν μεγάλα (πλοῖα), δεῖ πιστεύειν οἷς ἤδη ἔχομεν πλοίοις.

8. πολλὰ καὶ καλὰ δῶρά ποτε πρὸς ἐμὲ ἔπεμπες ἀνθ' ὧν νῦν σε λύσω.

Ex. 23. 1. Since you have broken the truce thus, we are not willing to march with you into the land of the barbarians.

2. In addition to this the general spoke as follows: I did not break the truce; but the enemy marched into the village and are now beating those in the market-place.

3. You ought not to say these things; for it is not lawful to insult the judge.

4. In the meanwhile they marched from the village along the road which leads across the plain towards the river.

5. For a long time we ruled (*Rule II, p.* 4) both land and sea very justly but afterwards the foreigners were stronger than we.

6. *You* say that this (man) is useful but *we* consider that the rest are more worthy.

REDUPLICATION

(1) Aspirates (θ, φ, χ) reduplicate with τ, π, κ; ῥ is doubled: *χορεύ-ω, κε-χόρευ-κα; θύ-ω, τέ-θυ-κα; ἐρρύηκα.*

(2) Verbs beginning with a vowel lengthen the vowel (as for Augment); this is retained in all moods. *ἀρτύω, ἤρτυκα, infin. ἠρτυκέναι.*

(3) Verbs beginning with two consonants (the second not being λ, ν, ρ) only *prefix* ε: *στρατεύω, perf. ἐστράτευκα; ψαύω, ἔψαυκα.*

In these verbs the pluperfect has ε only: *ἐψαύκη* 'I had touched.'

VERBS WITH STEMS ENDING IN -α, -ε, -ο

These form their tenses with a lengthened vowel. **α** and ε = η, ο = ω.

νικά-ω 'conquer' *νική-σω, ἐ-νίκη-σα, νε-νίκη-κα.*

ἀδικέ-ω 'injure' *ἀδική-σω, ἠδίκη-σα, ἠδίκη-κα.*

ζημιό-ω 'punish' *ζημιώ-σω, ἐ-ζημίω-σα, ἐ-ζημίω-κα.*

But verbs where **α** is preceded by ε, ι, ρ retain the **α**, as *δράω* 'act' *δρά-σω, ἔδρα-σα* (*Rule III, p.* 6).

THE VERB ἔχω

(1) The imperfect of *ἔχω* is *εἶχον* and this is used also as aorist, 'I had.'

(2) *ἔχω* with an adverb = to be in a certain state: *ἀθύμως ἔχει* 'he is despondent'; *ταῦτα οὕτως ἔχει* 'that is so.'

(3) Followed by an infin. or indirect question, *ἔχω* with a negative = to be able, to know: *οὐκ ἔχομεν λέγειν* 'we cannot tell.'

οὐκ ἔχει τί χρὴ λέγειν 'He does not know what he ought to say.'

AUGMENT IN COMPOUND VERBS

If the preposition ends with a vowel it is elided before the augment (except *περί* and *πρό*): *ἐκ* becomes *ἐξ*. Examples:

ἀπ-έ-πεμπ-ον, περι-έ-κρου-σα, ἐξ-έ-λυ-σα.

COMPOUNDS OF οὗτος (= 'tantus, tot, talis.')

τοσοῦτος, τοσαύτη, τοσοῦτο 'so great' (plural 'so many').

τοιοῦτος, τοιαύτη, τοιοῦτο 'such,' and the corresponding forms *τοσόσδε* and *τοιόσδε* have as relatives *ὅσος* and *οἷος*:

εἴχομεν τοσούτους στρατιώτας ὅσους οἱ βάρβαροι (εἶχον)
We had as many soldiers as the foreigners (had).

οἷος δεσπότης τοιοῦτος δοῦλος = like master, like man.

τὸ ἔργον deed, work.
ὁ φίλος friend.
ποιέω make, do, cause.
ἀγγέλλω announce.
περί *prep. with acc.* around, about; *with gen.* about = concerning.

Ex. 24. 1. ἆρα μὴ ἐθέλεις κακὰ λέγειν τὸν ἀγαθὸν κριτὴν ὃς οὐδέν σε κακὸν ἐποίησεν;

2. ἐπεὶ ἤκουσε ταῦτα, ὁ μέγας Ἀλέξανδρος πέμπει πρὸς τὸν Δαρεῖον τήνδε τὴν ἐπιστολήν· Ἐπεὶ μάχῃ νενίκηκα, πρῶτον μὲν τοὺς στρατηγούς σου, ἔπειτα δὲ ἐν τῷ νῦν πολέμῳ καὶ σὲ καὶ τὸν μετὰ σοῦ στρατόν, ἤδη ἔχω τὴν χώραν σου καὶ πρὸς τούτοις ἄρχω τῶν ἐν τῇ Ἀσίᾳ πολιτῶν. ἐπεὶ οὖν ταῦτα οὕτως ἔχει, ἆρ' οὐ δεῖ σὲ νομίζειν ἐμὲ εἶναι δεσπότην τοσούτων ὅσων σὺ πρὸ τῆς μάχης ταύτης ἦρχες, καὶ πιστεύειν ἐμοί; ἀγγέλλουσι γάρ σε κακὰ λέγειν περὶ ἐμοῦ καὶ περὶ τῶν Μακεδόνων. τοιαῦτα οὐκ ἔξεστί σοι. ζημιώσω οὖν καὶ σὲ καὶ τοὺς ἄλλους Πέρσας ἀνθ' ὧν ἠδικήκατε οὕτως ἡμᾶς ἢ λόγῳ ἢ ἔργῳ.

3. τῇ δὲ τρίτῃ ἡμέρᾳ ἐλύσαμεν τοὺς δούλους οὓς πιστοτάτους εἴχομεν ἐν τῷ χωρίῳ· οἱ γὰρ πολέμιοι μετὰ μεγάλου στρατοῦ ἐστρατεύκεσαν εἰς τήνδε τὴν χώραν καὶ ἤδη οὐ πολὺ ἀπεῖχον.

4. οἱ περὶ τὸν Κῦρόν (C. and his staff) ποτε πρὸ τῆς μάχης ἀθύμως εἶχον διὰ τοιάδε· οἱ μὲν γὰρ πολέμιοι ἦσαν πολλοί, ὁ δὲ στρατὸς οὗ Κῦρος ἦρχεν οὐ μέγας ἦν. ὁ δὲ Κῦρος ἔφη τάδε· Ὦ φίλοι, οὐ χρὴ ὑμᾶς ἀθύμως ἔχειν περὶ ἐκείνων. ὑμεῖς μὲν γὰρ Ἕλληνές (Greeks) ἐστε, ἐκεῖνοι δὲ οὐδὲν ἄλλο εἰσὶν ἢ βάρβαροι· διὸ νομίζω ὑμᾶς πολλῷ χρησιμωτέρους ἐν μάχῃ καὶ πολλῶν βαρβάρων.

Ex. 25. 1. "We shall certainly (*p.* 6 (2)) punish these (men)," said the general, "who have not only injured many of us but are sending out arms to the enemy."

2. He said that he was friendly to the people but for a long time he was sending letters concerning us to the general of the Persians.

3. It is not permitted to you to bribe these judges.

4. Nothing will now prevent the lawless-element (*neut. of adj.*) of the people from driving the faithful citizens out of this land.

5. When he had conquered (*aorist*)[1] that country, he ordered the judges to rule the people well.

[1] Except for some special reason use aorist for English pluperfect in dependent clauses containing a single verb.

THE PASSIVE VOICE

Perfect Passive		*Pluperfect Passive*	
I have been loosed		I had been loosed	
λέ-λυ-μαι	λε-λύ-μεθα	ἐ-λε-λύ-μην	ἐ-λε-λύ-μεθα
λέ-λυ-σαι	λέ-λυ-σθε	ἐ-λέ-λυ-σο	ἐ-λέ-λυ-σθε
λέ-λυ-ται	λέ-λυ-νται	ἐ-λέ-λυ-το	ἐ-λέ-λυ-ντο

Infin. λε-λύ-σ-θαι.

The perfect and pluperfect are given first as they show the personal endings best. In the other tenses the 2nd person singular is contracted since σ drops out between two hard vowels (α ε ο). Thus λυ-ε-σαι becomes λύ-ει or λύ-ῃ and ἐ-λυ-ε-σο = ἐλύου.

Present Passive		*Imperfect Passive*	
I am (being) loosed		I was being loosed (used to be loosed)	
λύ-ο-μαι	λυ-ό-μεθα	ἐ-λυ-ό-μην	ἐ-λυ-ό-μεθα
λύ-ει or λύ-ῃ	λύ-ε-σθε	ἐ-λύ-ου	ἐ-λύ-ε-σθε
λύ-ε-ται	λύ-ο-νται	ἐ-λύ-ετο	ἐ-λύ-ο-ντο

Infin. λύ-ε-σθαι.

Tenses ending in -μαι form the participle in -μενος, -η, -ον.

Present Part. λυ-ό-μενος. *Perf. Part.* λε-λυ-μένος.

The instrument and the means are expressed by the dative, the agent by ὑπό with the genitive case:

κεκώλυνται ὑπὸ τῶν πολεμίων
They have been hindered by the enemy.

ἐλελύμεθα ταῖς σπονδαῖς
We had been freed by the treaty.

The perfect denotes a state of things; the aorist an event.

CONSECUTIVE CLAUSES

The conjunction ὥστε with the infinitive expresses the natural or probable result of an action (negative μή). Used with the indicative it emphasises the actual result (negative οὐ).

(1) Are you wise enough to train others?
ἆρα σὺ οὕτω σόφος εἶ ὥστε ἄλλους παιδεύειν;

(2) So great was the storm that they refused to march
τοσοῦτος ἦν ὁ χειμὼν ὥστε οὐκ ἔφασαν στρατεύσειν.

ὥστε may also be used for 'therefore, and so' (Latin 'itaque').
μάχη ἐστὶν ἐν τῇ κώμῃ· ὥστε αἱ σπονδαὶ ἤδη λέλυνται.

παύω check, stop (*trans.*).
ἡσυχάζω remain quiet.
δια-βαίνω cross (over) (*trans.*).
ἐν ᾧ, ἐν ὅσῳ while.
παρα-σκευάζω prepare (*trans.*).
παρὰ (τὸν) νόμον contrary to law.
κατὰ (τὸν) νόμον according to law.
φέρ-ω bear, carry.

ὑπό *prep. with gen.* by (a person), *with acc. or dative*, under.

Ex. 26. 1. αἱ σπονδαί, ἔφη, λέλυνται. 2. ἔφη τὰς σπονδὰς λελύσθαι. 3. λέγει ὅτι πέπαυται ἡ μάχη. 4. οἱ πολέμιοι ἤδη ἐ-νε-νίκη-ντο. 5. οἱ τοξόται ἐκωλύοντο ὑπὸ τῶν πολιτῶν. 6. ἡ μεγάλη νίκη ἠγγέλλετο τῷ δήμῳ. 7. ὁ δὲ στράτος ἡσύχαζεν ἐν τῇ ὁδῷ, ὁ γὰρ ποταμὸς οὕτως ἰσχυρὸς ἦν ὥστε κωλύειν ἡμᾶς διαβαίνειν. 8. οἱ κριταὶ λέγονται εἶναι σόφοι. 9. τὰ δῶρα ἐλέγετο εἶναι καλά. 10. τὸ ἔργον εὖ πεποίηται. 11. ἆρ' οὐ παρασκευάζεται αὕτη ἡ οἰκία τῷ στρατηγῷ; 12. μετὰ δὲ ταῦτα, ἐν ᾧ ἔνιοι τῶν πολιτῶν, εὖ πεπαιδευμένοι ὑπὸ τοῦ στρατηγοῦ, παρεσκεύαζον τὰ πλοῖα πρὸς τὴν ναυμαχίαν, οἱ ἄλλοι ἐστράτευσαν κατὰ τὴν ὁδὸν ᾗ ἄγει ἀπ' Ἀθηνῶν πρὸς τὴν θάλατταν. 13. πολλὰ καὶ αἰσχρὰ λέγεται περὶ σοῦ ἀλλ' ἐγὼ τούτοις οὐ πεπίστευκα. 14. οἱ κλέπται, λελυμένοι ὑπὸ τῶν φίλων, ἔφευγον ἐκ τοῦ χωρίου. 15. τούτους δεῖν ἀξίως νομίζεσθαι κλέπτας λέγομεν, ἐπεὶ ἦσαν ἐν τῇ οἰκίᾳ καὶ ἀπ-έ-φερ-ον τὰ τοῦ κριτοῦ· ὥστε ζημιώσομεν τούτους διὰ τὰ κακὰ ἔργα ἃ πεποιήκασι παρὰ τοὺς νόμους.

Ex. 27. 1. While these things were being announced, the people (*singular*) were keeping quiet. 2. He said (that he) had been injured by the disgraceful words of the poet. 3. The army has been prevented from crossing the river; we will therefore march towards the enemies' country which is distant a journey (*acc.*) of not many days. 4. The treaty had already been disgracefully broken by the Persians; and so ('itaque') nothing prevented us from marching into that land. 5. The words of the general are heard both by the army and by the people. 6. While some were being loosed by the soldiers, others were being armed by the general. 7. According to the poet great deeds are worthy of beautiful words. 8. Not having been loosed by their friends they were despondent, naturally (= πῶς γὰρ οὔ;). 9. Are you clever enough to make a table (οὕτως...ὥστε)? 10. Those islands which we once ruled are said to have been conquered by the barbarians. 11. It is said that[1] the fight has been stopped by the police.

[1] Either "The fight is said..." or "they say that" or λόγος ἐστί and infin.

USES OF *αὐτός*

αὐτ-ός -ή -ό = 'ipse' = self.

(1) In apposition to nouns and pronouns *αὐτός* is used to mean 'self.'

(*ἐγὼ*) *αὐτὸς ἐποίησα* 'I did it myself'—'ipse feci.'

ἔφασαν αὐτὸν τὸν κριτὴν ποιῆσαι = They said that the judge himself did it.

ἡ κόρη ταῦτα ἔφη αὐτή = The girl said this herself.

(2) As an adjective *ὁ αὐτός* = the same. Where the vowels clash it is often combined with the article (*κρᾶσις* = mixing). It is followed by a dative.

Nom.	*ὁ αὐτός* (*αὐτός*)	*ἡ αὐτή* (*αὐτή*)	*τὸ αὐτό* (*ταὐτό*)
Acc.	*τὸν αὐτόν*	*τὴν αὐτήν*	*τὸ αὐτό* (*ταὐτό*)
Gen.	*τοῦ αὐτοῦ* (*ταὐτοῦ*)	*τῆς αὐτῆς*	*τοῦ αὐτοῦ* (*ταὐτοῦ*)
Dat.	*τῷ αὐτῷ* (*ταὐτῷ*)	*τῇ αὐτῃ* (*ταὐτῇ*)	*τῷ αὐτῷ* (*ταὐτῷ*)

So in the nom. plural

οἱ αὐτοί (*αὑτοί*) *αἱ αὐταί* (*αὑταί*) *τὰ αὐτα* (*ταὐτά*)

but in the other cases there can be no 'crasis.'

Further examples of 'crasis': *κἀγώ* = *καὶ ἐγώ*, *τἀνθρώπου* = *τοῦ ἀνθρώπου*.

The various forms of *αὐτός* must be carefully distinguished from those of *οὗτος*. Note that *αὐτός* always has the accent on the final syllable, *οὗτος* never.

αὐτὴ ἐποίησεν	*αὕτη ἐποίησεν*	*αὑτὴ ἐποίησεν*
She did it herself	this woman did it	the same woman did it

ταῦτα = these things. *ταὐτά* = the same things.

ὁ κριτὴς αὐτός = the judge himself.

ὁ αὐτὸς κριτής = the same judge.

οὗτος ὁ κριτής = that judge.

N.B. All cases of *αὐτός*, except the nominative, are used for 'him,' 'her,' 'it.' Latin 'eum,' 'eam,' 'id'; *αὐτόν*, *αὐτήν*, *αὐτό*. 'I shall punish him' *ζημιώσω αὐτόν*. 'We do not trust them' *οὐ πιστεύομεν αὐτοῖς*. *ἡ οἰκία αὐτοῦ* 'his house.'

Aorist Passive		*Future Passive*	
I was loosed		I shall be loosed	
ἐ-λύ-θην	*ἐ-λύ-θημεν*	*λυ-θή-σομαι*	*λυ-θη-σόμεθα*
ἐ-λύ-θης	*ἐ-λύ-θητε*	*λυ-θή-σει* (or *-ῃ*)	*λυ-θή-σεσθε*
ἐ-λύ-θη	*ἐ-λύ-θησαν*	*λυ-θή-σεται*	*λυ-θή-σονται*
Infin. *λυ-θῆ-ναι.*		*Infin.* *λυ-θή-σεσθαι.*	

εὐθύς } immediately.
αὐτίκα }
προσ-βάλλω attack (*dat.*).
ἀποθνήσκω die, am killed.
ἡ νόσος disease.
ἐλπίζω hope, expect (*usually future infin.*).
ὁ σεισμός earthquake.

The following verbs form aorist pass. in -σθην, fut. passive -σθήσομαι: κλείω 'close,' σείω 'shake,' ἀκούω 'hear,' κελεύω 'order.' They also form perfect pass. and pluperfect pass. -σμαι, -σμην.

Ex. 28. 1. ὁ μὲν οὖν ταῦτα ἔπειθεν· οἱ δὲ ἄλλοι ταὐτὰ ἔλεγον.
2. τῇ δὲ αὐτῇ ἡμέρᾳ ἦν σεισμὸς γῆς ἐν ταύτῃ τῇ νήσῳ.
3. πῶς ἐλπίζεις νικήσειν ἐπεὶ ὀλίγους μὲν ἔχεις στρατιώτας οἱ δὲ πολέμιοι τοσοῦτοί εἰσιν;
4. μετὰ ταῦτα ἐν ὅσῳ προσεβάλλομεν τοῖς Πέρσαις, αὐτίκα ἐκελεύσθην ὑπ' αὐτοῦ τοῦ στρατηγοῦ ἄγειν τοὺς ἐφ' ἵππων κατὰ τὴν πρὸς Ἀθήνας ὁδόν.
5. ἐπεὶ αἱ θύραι οὐκ ἐκλείσθησαν ἐκέλευσεν αὐτὰς κλεισθῆναι εὐθύς.
6. τῶν κλεπτῶν ἄλλοι ἐλύθησαν, ἄλλοι ζημιωθήσονται.
7. ἐν ᾧ τοσοῦτοι τῶν στρατιωτῶν νόσῳ ἀπέθνησκον, ἀδύνατον ἦν προσβάλλειν τοῖς ἐν τῷ χωρίῳ.
8. ἐκέλευσεν αὐτὸν λυθῆναι αὐτίκα ἐπεὶ οὐδὲν κακὸν ἐποίησεν.
9. ἀγγέλλουσιν ὅτι ὁ μὲν στρατὸς ἐνικήθη οἱ δὲ βάρβαροι ἤδη διαβαίνουσι τὸν ποταμόν.
10. ἐπεὶ οὐκ ἐπίστευσα τούτοις τοῖς δούλοις, αὐτὸς ἔλυσα τὸν ἵππον.
11. ἡ θύρα ἤδη κέκλεισται· τί οὖν δεῖ ἡμᾶς ποιεῖν;
12. ἐν ταύτῃ ἐπιστολῇ εἰσὶ καὶ ἀγαθοὶ καὶ κακοὶ λόγοι.

Ex. 29. 1. They say that this victory is great; and that letter announces the same things. 2. On the same day we attacked Cyrus' army since we were hoping to defeat him. 3. These young men, having been well trained by the judge, are being sent to the general. 4. Since these (men) have done nothing disgraceful they will be set free at once. 5. Is it not disgraceful to have been conquered by barbarians, and that (too) by a few (men)? 6. *You* are allowed to keep quiet in the village but I must march into the land of the Persians. 7. They said that those thieves would be set free by the police themselves. 8. While the soldiers were preparing their arms, he ordered them to attack immediately. 9. Since so-many were dying of disease in that war, the army was ordered to flee from the land. 10. He said that many houses had been destroyed (δια-λύω) by that earthquake.

(1) Note the difference between:

ὁ πολίτης ἔφη αὐτὸς ποιῆσαι
The citizen said that he did (it) himself.

ὁ πολίτης ἔφη αὐτὸν ποιῆσαι
The citizen said that he (some one else) did it.

ὁ πολίτης οὐκ ἔφη αὐτὸς ἀλλὰ τὸν δοῦλον ποιῆσαι
The citizen said that not he but the slave did it.

The difficulty in English is that we have only one pronoun for the 3rd person.

(2) *αὐτ-όν, -ήν, -ό* are often omitted when easily supplied; as above.

(3) There is a double reflexive pronoun (reciprocal) *ἀλλήλ-ους -ας -α* which has no singular and no nominative, and corresponds to 'each other' or 'one another' *e.g.* *πιστεύουσιν ἀλλήλοις* 'they trust each other.'

(4) Note the phrase:

He was too wise to trust the slave
σοφώτερος ἦν ἢ ὥστε πιστεύειν τῷ δούλῳ.

Latin: 'sapientior erat quam ut' etc.

IRREGULAR COMPARISON

μέσος	middle	*μεσαίτερος*	*μεσαίτατος*
ἴσος	equal	*ἰσαίτερος*	*ἰσαίτατος*
φίλος	beloved	*φιλαίτερος*	*φιλαίτατος*
		and	*φίλτατος*
ἥσυχος	quiet	*ἡσυχαίτερος*	*ἡσυχαίτατος*
πλησίον	*adv.* near	*πλησιαίτερον*	*πλησιαίτατα*

N.B. *οἱ πλησίον* 'one's neighbours.' The adverb *ἴσως* often = perhaps.

POSSESSOR

The genitive of pronouns is often used instead of the possessive adjective (like 'eius,' 'eorum' in Latin), *e.g.* *ἡ οἰκία μου* (not *ἐμοῦ*) 'my house.' So also *σου, ἡμῶν, ὑμῶν* and *αὐτοῦ*, but if reflexive, *ἑαυτοῦ* must be used for *αὐτοῦ* and placed between the article and the noun.

'We conquered his army' *ἐνικήσαμεν τὸν στρατὸν αὐτοῦ.*

'He set free his (own) slaves' *ἔλυσε τοὺς ἑαυτοῦ δούλους.*

ἤ...ἤ either...or.
τολμάω dare.
ὑπ-οπτεύω suspect.
διότι, ὅτι because.
πλησίον *adv.* near, neighbouring.
διώκω pursue.
ἀεί always.
οὐκέτι no longer, no more.
ἐπί *prep. with acc.* on to, against; *with gen. or dat.* on.

Ex. 30. 1. οὐχ ὑπώπτευσά σε ταῦτα ποιῆσαι· δεῖ γὰρ τοὺς φίλους ἀεὶ πιστεύειν ἀλλήλοις.

2. νομίζω οὖν ὅτι ὁ κλέπτης ζημιωθήσεται· οἱ γὰρ δικασταὶ δικαιότεροί εἰσιν ἢ ὥστε πείθεσθαι τοῖς δώροις αὐτοῦ.

3. οὐκ ἔφη αὐτὸς ἀδικηθῆναι ἀλλὰ τὸν ἑαυτοῦ φίλον.

4. μετὰ δὲ ταῦτα αἱ σπονδαὶ ἐλύθησαν, ὅτι οἱ Ἀθηναῖοι καὶ οἱ Λακεδαιμόνιοι οὐκέτι ἐπίστευον ἀλλήλοις· οὗτοι μὲν γὰρ ὥπλιζον πολλοὺς στρατιώτας, ἐκεῖνοι δὲ παρεσκεύαζον τὰ ἑαυτῶν πλοῖα εἰς μάχην.

5. ἐν μὲν τῇ ἀρχῇ τοῦ πολέμου συνεκλείσθημεν ἐν ταῖς Ἀθηναῖς, ὕστερον δέ, ἐπεὶ ἐν τῇ ναυμαχίᾳ ἐνίκησαν οἱ ναῦται ἡμῶν, ἐτολμήσαμεν προσβάλλειν τοῖς Πέρσαις ἐν μέσῳ τῷ πεδίῳ.

6. ἆρ' οὐκ ἀγαθὸν τοὺς φίλους εὖ ποιεῖν ἀλλήλους, οὐχ ὅτι δεῖ τοῦτο ποιεῖν ἀλλὰ διότι πιστοί εἰσιν ἀλλήλοις ἀεί;

7. τί τοὺς πολεμίους ἤδη νενικημένους οὐ διώκεις, ὦ στρατηγέ;

8. ἆρα μὴ ἐτόλμησας ὑποπτεύειν τὸν στρατηγόν σου περὶ τούτων, ὃς οὐ μόνον πολλάκις νενίκηκε τοὺς πολεμίους ἀλλὰ καὶ εὖ πεποίηκε καὶ σὲ καὶ τοὺς πλησίον σου;

Ex. 31. 1. After that since he suspected that the enemy would march towards us he dared to cross the river himself while they were preparing their arms. But in the meanwhile another great army began-to-attack and so we were defeated. 2. Thieves are said to trust one another but I do not believe that they do so (that) always. 3. He is too wise to be without arms on that road because travellers (Ex. 20) are often pursued by highwaymen in our country. 4. Some say one thing and some say another but the judge thinks that these ships are being prepared contrary to the law. 5. Our house is nearer to the sea and much quieter than those in the village. 6. He said that he would not dare to do such things himself. 7. I have always suspected that those men were preparing arms for the barbarians. 8. Since they have injured our friends, we must no longer trust them. 9. The slave says that he shut the door; how (is it) then (that) the horse is in the road? Perhaps the horse itself dared to open (ἀνοίγω) the stable-door (Ex. 18, No. 11).

THE MIDDLE VOICE

Its formation is the same as that of the Passive except in the future and aorist.

Fut. Middle		*Aorist Middle*	
I shall loose for myself		I loosed for myself	
λύ-σομαι	*λυ-σόμεθα*	*ἐ-λυ-σάμην*	*ἐ-λυ-σάμεθα*
λύ-σει or *-σῃ*	*λύ-σεσθε*	*ἐ-λύ-σω*	*ἐ-λύ-σασθε*
λύ-σεται	*λύ-σονται*	*ἐ-λύ-σατο*	*ἐ-λύ-σαντο*
Infin. *λύσεσθαι.*		*Infin.* *λύσασθαι.*	

The meaning of the Middle Voice of a verb can best be learnt from examples.

(1) To do something for one's own advantage.
(2) To get something done for one's own advantage.
(3) To do something to oneself (a few verbs only). *E.g.*

(1) *κομίζω* 'I convey' *Mid.* *κομίζομαι* 'I take away for myself.'
σώζω 'I save' *Mid.* *σώζομαι* 'I save (what is my own).'
(2) *διδάσκω* 'I teach' *Mid.* *διδάσκομαι τὸν υἱόν* 'I have my son taught.'
ᾠκοδόμησα 'I built' *Mid.* *ᾠκοδομησάμην* 'I had a house built.'
(3) *λούω* 'I wash' (*trans.*) *Mid.* *λούομαι* 'I bathe.'
κόπτω 'I beat' *Mid.* *κόπτομαι* 'I beat myself, I mourn.'

Some verbs are used only in the middle voice and are called Middle Deponents (p. 32), some in the passive only and are called Passive Deponents. It is always easy to distinguish *λύομαι* Middle from *λύομαι* Passive, because the Middle is followed by an object (acc., gen., or dat.).

Some verbs change their meaning considerably in the Middle Voice and the case of the object is often different:

πείθω	I persuade (*acc.*)	*Mid.* *πείθομαι*	I obey (*dat.*)
παύω	I check (*acc.*)	*Mid.* *παύομαι*	I cease from (*gen.*)
λύω	I set free (*acc.*)	*Mid.* *λύομαι*	I ransom (*acc.*)

Many verbs that express some very personal action have a fut. mid. instead of a fut. act. They are very often intransitive; *e.g.*

φεύγω 'I flee' *φεύξομαι* *ἀκούω* 'I hear' *ἀκούσομαι*
διώκω 'I pursue' *διώξομαι*

The future of *εἰμί* ('sum') is as follows:

ἔσ-ομαι	*ἐσ-όμεθα*	*Partic.* *ἐσ-όμενος -η -ον*
ἔσ-ει or *-ῃ*	*ἔσ-εσθε*	*Infin.* *ἔσ-εσθαι*
ἔσ-ται	*ἔσ-ονται*	

ὁ χρυσός gold.
ἡ ἀρετή virtue, courage.
*τε = both (with καί).
ἡ σοφία wisdom.
*μέντοι however.
καίτοι (and) yet.

ἐποίησε τράπεζαν	he made a table.
ἐποίησε πόλεμον	he caused a war.
ἐποιήσαντο σπονδάς	they made a treaty.

περὶ πολλοῦ (ὀλίγου etc.) ποιεῖσθαι 'to consider of great (little) importance.'

Ex. 32. 1. οἱ μέντοι στρατιῶται οὐκ ἤθελον πείθεσθαι τῷ στρατηγῷ.
2. οἱ οὖν πολῖται πολλῷ χρυσῷ ἐλύσαντο τὸν στρατηγόν, οἱ γὰρ πολέμιοι οὐκ ἤθελον λύειν αὐτόν.
3. μετὰ δὲ ταῦτα ἐπαύσαντο τοῦ πολέμου οἵ τε Λακεδαιμόνιοι καὶ οἱ Ἀθηναῖοι καὶ σπονδὰς ἐποιήσαντο ἐπὶ τοῖσδε (*p.* 20).
4. μετὰ τὸν πόλεμον οἰκοδομήσομαι μεγάλην οἰκίαν πλησίον τῆς θαλάττης, καὶ ἐν αὐτῇ ἔσται πολλὰ καὶ καλά, ἐπεὶ οὐκ ὀλίγος χρυσὸς ἔστι μοι.
5. ὁ μὲν οὖν ταῦτα ἔπειθεν· οἱ δὲ πολῖται οὐκ ἐπείθοντο αὐτῷ.
6. ἐπεὶ πρῶτον ταῦτα ἀκούσονται, οἱ νεανίαι αὐτίκα ἀποφεύξονται.
7. ἔφη αὐτὸς περὶ πολλοῦ ποιεῖσθαι τὴν σοφίαν, τὸν δὲ δῆμον ἀεὶ νομίζειν τὸν χρυσὸν εἶναι χρησιμώτερον τῆς ἀρετῆς.
8. τί οὐκ ἐπαύσω τῶν κακῶν λόγων, ὦ πολῖτα; ἔφασαν οἱ κριταί· ἐπεὶ οὖν οὐκ ἤθελες ἡσυχάζειν, ἢ ζημιωθήσει πολλῷ χρυσῷ ἢ κατα-κλεισθήσει δέκα ἡμέρας ἐν τῷ δεσμωτηρίῳ.
9. ἐν δὲ τούτῳ οἱ περὶ τὸν Κῦρον ὡπλίζοντο καὶ παρεσκευάζοντο πρὸς μάχην ὅτι ἐνόμιζον τοὺς Πέρσας πλησίον ἔσεσθαι ταύτῃ ἡμέρᾳ.

Ex. 33. 1. Shall we not cease from war, and make a treaty upon just (terms)?
2. Sōcratēs himself attached-great-importance-to wisdom and yet the Athenians were not often persuaded by him.
3. Why, o Charmides, did you not ransom your friend? Do you not consider that a faithful friend is more useful to you than much gold?
4. It is said that both the Lacedaemonians and the Athenians have ceased from war and made a treaty; however those in the islands do not believe that these things are so.
5. Of the enemy some were arming (*mid.*), some were already attacking, others were fleeing since they thought (*imperf.*) that their friends had been conquered.
6. Do you expect (ἐλπίζω) others to stop this war which you yourself caused?

DEPONENT VERBS

βούλομαι	I desire	(*infin.*)	*ἕπομαι* accompany, follow (*dat.*).
πειράομαι	try, attempt	(*infin.*)	*μάχομαι* fight (*dat.*).
δύναμαι	am able, can	(*infin.*)	*μέμνημαι* remember (*gen.*).
ἐπίσταμαι	know (how)	(*infin.*)	*πορεύομαι* march, travel.

Verbs in *-αμαι* are conjugated like *λέλυμαι.*

Present		*Imperf.*	
δύνα-μαι	*δυνά-μεθα*	*ἐ-δυνά-μην*	*ἐ-δυνά-μεθα*
δύνα-σαι (*δύνῃ*)	*δύνα-σθε*	*ἐ-δύνα-σο* (*ἐδύνω*)	*ἐ-δύνα-σθε*
δύνα-ται	*δύνα-νται*	*ἐ-δύνα-το*	*ἐ-δύνα-ντο*

Infin. *δύνα-σθαι.* *Partic.* *δυνά-μενος.*

POSSESSIVE ADJECTIVES

ἐμ-ός -ή -όν	my, mine	*ἡμέτερ-ος -α -ον*	our, ours
σός σή σόν	thy, thine	*ὑμέτερ-ος -α ον*	your, yours

These like other adjectives must be preceded by *ὁ, ἡ, τό, e.g.*

ἡ ἐμὴ οἰκία = *ἡ οἰκία μου* = my house.
ἡ ὑμετέρα πόλις = *ἡ πόλις ὑμῶν* = your city.
οἱ σοὶ δοῦλοί εἰσι πολλῷ πιστότεροι τῶν ἐμῶν.
Your slaves are far more faithful than mine.

THE INFINITIVE

Besides the uses of the infin. already met with it can be used (1) like the supine in *-u* (Latin) to define certain adjectives, *e.g.*

δεινὸς λέγειν = clever at speaking;
ἄξιος ἀκούειν = worth hearing;

(2) to take the place of the gerund of the verb when a neuter article *τό* is put before it and declined:

Nom. and Acc.	*τὸ ἀκούειν*	hearing	('audiendum')
Gen.	*τοῦ ἀκούειν*	of hearing	('audiendi')
Dat.	*τῷ ἀκούειν*	by hearing	('audiendo').

This often takes the place of a noun after a preposition: *ἀντὶ τοῦ παρασκευάζειν* = instead of preparing. Adverbs etc. are placed between the article and the infinitive, with the negative, which, if required, will be *μή.*

διὰ τὸ μὴ αὐτοὺς εὖ παιδευθῆναι 'on account of their not being well trained' or 'due to their lack of education.'

κάθημαι I sit, *imperf.* ἐκαθήμην.
(κατα-)κεῖμαι I lie (down), *imperf.* ἐκείμην.
ἀντί *prep.* instead of (*gen.*).
ὁ ἰα-τρός physician.
λοιπός -ή -όν left, remaining.
πάρ-ειμι I am present, arrive.
ὁ σῖτος corn, food, *pl.* σῖτα.

Ex. 34. 1. τί οὐ βούλεσθε μάχεσθαι, ὦ νεανίαι; ἑπόμεθα γὰρ στρατηγῷ ὃς ἐπίσταται τὰ τοῦ πολέμου· δεῖ οὖν τοιούτῳ πιστεύειν.

2. οἱ οὖν ἡμέτεροι στρατιῶται οὐκ ἐδύναντο διαβαίνειν τὸν ποταμόν· οὐ γὰρ εἶχον πλοῖα.

3. ἆρ' οὐκ ἔδει σε, ὦ νεανία, μεμνημένον τῶν τοῦ ἰατροῦ λόγων, ἡσυχάζειν ἐν τῇ οἰκίᾳ ἀντὶ τοῦ ἱππεύειν καὶ τοῦ χορεύειν;

4. ἐγώ ποτε καὶ οἱ δοῦλοί μου, πορευόμενοι δι' ἐρημοτάτου πεδίου, μετὰ πολλὰς ἡμέρας ἀπεθνήσκομεν διὰ τὸ μὴ σῖτα ἔχειν. οἱ μέντοι φίλοι ἑπόμενοι οὐ μετὰ πολὺν χρόνον παρ-ῆσαν καὶ ἐδύναντο σώζειν ἡμᾶς ὥστε μὴ ἀποθνήσκειν.

5. ἡ ἡμετέρα οἰκία κεῖται πρὸς τῇ ὁδῷ.

A sorry plight.

νεανίας ποτὲ ὃς ἀθύμως εἶχεν διὰ δεινὴν νόσον, ἔφη τῷ ἰατρῷ τοσαύτην νόσον ἔχειν ὥστε μὴ δύνασθαι μήτε καθῆσθαι μήτε κατακεῖσθαι μήτε ἑστάναι· ὁ δὲ ἰατρὸς ἔφη, Οὐδὲν ἄλλο σοι λοιπόν ἐστι, ὦ φίλε, ἢ κρέμασθαι.

Ex. 35. 1. The soldiers therefore refused to march through that very desolate land since so many were dying of disease on account of their not having food. 2. He says that he does not know how to ride and that he wishes on that account to remain quiet in the house. 3. Being unable to escape from prison, we were hoping that the citizens would either ransom us or would perhaps make a treaty. 4. Do you not remember the words of the wise judge who said that faithful citizens ought to obey the laws which (they) themselves made? 5. He said that he would try (*Rule III, p.* 6) to be present. 6. Instead of preventing us from fighting you ought to persuade the rest to arm-themselves (*mid.*). 7. They say that the judge is clever at speaking and worth hearing; I therefore, wishing to be present on that day, journeyed for many days from Athens to your village. He himself will be present and will educate the citizens about these laws 8. Some are sitting in the houses, others are lying in the market-place.

REVISION

1. Parse *ἦν, ἥν, ταῦτα, ταὐτά, ἄλλα, ἀλλά, λῦσαι, λύσει* (two ways).

2. Greek for: he is, he will be, ye are, he persuades, you obey, such things.

3. Perf. infin. act. of *ἱππεύω, χορεύω, πιστεύω, κλείω.*

4. Gen. sing. of door, sea, sailor, time, self, thou, youth.

5. Make a table classifying the mute consonants.

6. Greek for: in addition to this, meanwhile, on horseback, before the war, according to the law, after the battle, with us, on the table, without arms, by my words, against the enemy, by the judge, among the thieves.

7. Give the fut. and aorist of *ἀκούω, πορεύομαι, οἰκοδομέω, ζημιόω.*

8. Greek for: immediately, more usefully, we ourselves, once, the rest, much wiser, a great (thing), very wisely, than, nothing, however.

9. Form the imperfect of *ὁπλίζω, οἰκτείρω, μάχομαι, ἐκπέμπω, παρασκευάζω, ἔχω, ἄρχω, ὑβρίζω, ῥίπτω, πάρειμι.*

10. When should (1) *οὐχ* be used instead of *οὐ,*
(2) *μή* „ „ *οὐ,*
(3) *ἔστι* „ „ *εἰσί,*
(4) *δέ* „ „ *ἀλλά,*
(5) *ἔχω* „ „ *εἰμί,*
(6) *τάδε* „ „ *ταῦτα?*

11. Parse *ἐλύσω, ἐνενικήκη, ἦρε, φεύξει, ἦσθα, ἐσόμενον.*

12. Greek for: to sit, to have been loosed, he was checked, you shall be punished, he had been washed, to bathe.

13. Translate:
(*a*) Some say one thing, some say another.
(*b*) He said that he would do it himself (*three words*).
(*c*) I am not such a fool as to trust that man.
(*d*) Cyrus and his staff were despondent about this.
(*e*) We consider wisdom to be of great importance.
(*f*) Instead of fighting each other, you ought to be friends.
(*g*) He advised the islanders to cease from war.
(*h*) Why don't you trust the friends you have?

PART II

THE THIRD DECLENSION

Nouns in this declension have (1) a consonant stem, (2) a stem in soft vowels (*ι*, *υ*); a few have (3) a stem in -*ο* or -*ω*.

		Masc. and Fem.	*Neuters*
Singular.	*Nom.*	-ς	No ending.
	Voc.	—	No ending.
	Acc.	-α (vowel stems -ν)	No ending.
	Gen.	-ος (some vowel stems -ως)	-ος
	Dat.	-ι	-ι

		Masc. and Fem.	*Neuters*
Plural.	*Nom. and Voc.*	-ες	-α
	Acc.	-ας	-α
	Gen.	-ων	-ων
	Dat.	-σι	-σι

VOCATIVES

If a noun is one often used in the Vocative, this will be the stem of the noun after dropping those final consonants which may not stand at the end of a word.

A Greek word ends either in a vowel or in ν ρ ς (ψ and ξ). The only exceptions are words like ἐκ, οὐκ which go closely with other words.

Noun	*Stem*	*Vocative*	*English*
πατήρ	πατερ-	πάτερ	father
γέρων	γεροντ-	γέρον	old man
γυνή	γυναικ-	γύναι	woman
βασιλεύς	βασιλευ-	βασιλεῦ	king
παῖς	παιδ-	παῖ	boy

So with neuters the *Nom. Voc. Acc.* drop the unsuitable consonant or change it, in some, to ς.

κερατ- horn, nom. κέρας,
σωματ- body, nom. σῶμα.

THIRD DECLENSION

Stems ending in Gutturals (κ, γ, χ), cp. Latin 'leg'—nom. 'lex.' Nouns in this class are masc. or fem. only.

Verbs similarly formed—πλέκ-ω, fut. πλέξω, aor. ἔπλεξα.

Most of these verbs form pres. stem in -ττω (-σσω).

φυλακ- 'guard, sentinel.'			*Verb stem* φυλακ-	
	Sing.	*Plural*	*Pres.*	φυλάττω
N. V.	φύλαξ	φύλακ-ες	*Fut.*	φυλάξω
Acc.	φύλακ-α	φύλακ-ας	*Aor.*	ἐ-φύλαξα
Gen.	φύλακ-ος	φυλάκ-ων	*Perf.*	πε-φύλαχα
Dat.	φύλακ-ι	φύλαξι(ν)	*Perf. Pass.*	πε-φύλαγ-μαι
			Aor. Pass.	ἐ-φυλάχ-θην

Similarly, stems in Labials π, β, form nom. sing. ψ; γύψ (γυπ-) 'vulture,' dat. pl. γυψί.

For the conjugation of πε-φύλαγ-μαι see p. 40.

Some verbs have also a strong perfect active with an intransitive meaning: πράττω 'I do, I fare'; πέ-πραγ-α 'I have fared.' A strong tense is formed direct from the verb stem.

εὖ πέ-πραχα ταῦτα = I have done this well.
εὖ πέ-πραγ-α = I have fared well.

δή (in questions and answers often δῆτα), a strengthening particle placed after words to call special attention.

τί δή; τί δῆτα; = what pray? why pray? οὐ δῆτα = certainly not.
καὶ δή = and what is more.... καὶ δὴ καί = and especially.
τότε δή = ('tum vero') then indeed. σὺ δή... = you of all people....

Sometimes it can be translated 'of course, naturally, you know.' ῥᾳδίως δή 'easily as you might suppose.'

Note the following.

πῶς; how? οὕτως thus, οὐδαμῶς in no way, ῥᾳδίως easily.
ποῦ; where? ἐκεῖ there, οὐδαμοῦ nowhere, πανταχοῦ everywhere.
πότε; when? τότε then, οὐδέποτε never, ἄλλοτε at another time.
πόθεν; whence? ἐκεῖθεν thence, οὐδαμόθεν, ἄλλοθεν.
ποῖ; whither? ἐκεῖσε thither, οἴκαδε homewards, δεῦρο hither.

λέγ-ω I say, mean; λέξω, ἔλεξα.
κηρύττω proclaim, *stem* κηρυκ-.
οὐ...πω / οὐδέπω } not yet.
τάττω draw up, arrange, *stem* ταγ-; *mid.* τάττομαι, pay (tribute).
πράττω, do, fare, transact, exact (a penalty), *stem* πραγ-.
δέχ-ομαι (*deponent*) accept, receive; δέξομαι, ἐδεξάμην.
ταράττω trouble, confuse; *aor. pass.* ἐ-ταράχ-θην.
ἄρχ-ω rule (*gen.*); *mid.* ἄρχομαι, begin, *aor. part.* ἀρξάμενος.

ὁ κῆρυξ κηρυκ- herald.
ἡ φάλαγξ φαλαγγ- phalanx.
ὁ ὄνυξ ὀνυχ- nail, hoof.
ἡ γυνή γυναικ- woman, wife.
ἡ νύξ νυκτ- night.
τέλος *adv.* at last.

Ex. 36. 1. οἱ μὲν δὴ ταῦτα ἔ-πραξαν καὶ ἔλεξαν· ὁ δὲ Κῦρος ἐκέλευσεν αὐτοὺς πέμπεσθαι εἰς τὸ δεσμωτήριον μετὰ πολλῶν φυλάκων. 2. μετὰ δὲ ταῦτα πολλοὶ τῶν ἐν ταῖς νήσοις ἐτάξαντο χρυσὸν ἀντὶ τῶν πλοίων. 3. οὐκ ἔφη δέξεσθαι ἃ ἔπεμπον δῶρα. 4. τότε δὴ οἱ στρατιῶται οἴκαδε ἔφευγον καὶ δὴ καὶ οἱ τῆς φάλαγγος αὐτῆς ἐταράχθησαν. 5. κακῶς οὖν πεπράγαμεν, νενικημένοι δὴ καὶ κατὰ γῆν καὶ κατὰ θάλατταν. 6. τί πέπραχας, ὦ νεανία; σὺ δὴ ἐδέξω χρυσὸν παρὰ τῶν πολεμίων; 7. τῇ δὲ αὐτῇ νυκτὶ οἱ κήρυκες ἔλεξαν ὅτι ὁ τῶν πολεμίων στρατὸς οὐ πολὺ ἀπέχει. 8. ἆρα μὴ τούτῳ δὴ στρατηγῷ πιστεύσετε; ὃς ἐπίσταται οὔτε τάττειν τοὺς στρατιώτας πρὸς μάχην οὔτε πράττειν τὰ τοῦ πολέμου. 9. ποῦ δῆτά ἐστιν ὁ τοῦ δεσμωτηρίου φύλαξ; βουλόμεθα γὰρ λύεσθαι τὴν τοῦ κριτοῦ γυναῖκα. 10. παρὰ τούτους οὖν δεῖ ἡμᾶς μεγάλα πέμπειν δῶρα, πεφυλάχασι γὰρ πολὺν χρόνον τὰς ἡμετέρας γυναῖκας. 11. οὗτος δὴ λέγεται εὖ πεπραχέναι τὰ τῆς νήσου ἧς ἦρξεν. 12. φυλαττόμενοι καὶ καθ' ἡμέραν καὶ κατὰ νύκτα οὐ δύνανται οὐδέποτε ἐκφεύγειν. 13. τεταραγμένοι τοῖς τοῦ κήρυκος λόγοις πανταχόσε φεύγουσιν ἐκεῖθεν.

Ex. 37. 1. When he had (Ex. 25, No. 5) transacted this business, he ordered the army to be drawn-up at once on the road. 2. The same night many other things were being announced and especially that our troops at last had won a great victory. 3. Now (*p.* 11) Cyrus had not yet drawn up his army: some were sitting by the roadside, others were bathing in the river. 4. Since however the citizens did not ransom me, I was guarded in the prison by the sentinels. 5. (While) lying in the river the hippopotamus was disturbed by the crocodile. 6. Owing to the fact that (*p.* 32) the work was not yet finished (δια-πράττω) we were ordered not to cease from it by day or night. 7. "My boy (O boy), why, pray, did you open my letters?" "O father," said he, "I did not do (it)."

Note. 'Αθήναζε 'to Athens,' 'Αθήνησι 'at Athens,' 'Αθήνηθεν 'from Athens.' So also Θῆβαι 'Thebes,' Πλαταιαί 'Plataea,' 'Ολυμπία 'Olympia,' but apply *Rule III*, *p.* 6.

LABIAL VERBS (π, β, φ)

These form fut. in -ψω, aor. in -ψα, perf. in -φα.

γράφω 'I write,' stem γραφ-, fut. γράψω, aor. ἔγραψα, perf. γέγραφα, perf. pass. γέγραμμαι, aor. pass. ἐγράφθην.

(1) Many of these add τ to the present: βλάπτω 'I hurt.'

(2) Some have strong aor. pass. as well: ἐ-γράφ-ην.

DENTAL NOUNS AND VERBS (τ δ θ)

	Sing.	*Plural*	*Stem*
ὁ, ἡ	παῖς boy, girl.	παῖδ-ες	παιδ-
	παῖ	παῖδ-ες	
	παῖδ-α	παῖδ-ας	
	παιδ-ός	παίδ-ων	
	παιδ-ί	παι-σί	

Neuters drop τ in nom. sing.

N. V. A.	ὄνομα name.	ὀνόματ-α	ὀνοματ-
Gen.	ὀνόματ-ος	ὀνομάτ-ων	
Dat.	ὀνόματ-ι	ὀνόμα-σι	

The following have acc. sing. in -ιν: χάρις (χαριτ-) fem. 'favour'; ἔρις (ἐριδ-) fem. 'strife.' ὄρνις (ὀρνιθ-) 'bird,' ὁ or ἡ, often has acc. sing. ὄρνιν, acc. pl. ὄρνεις.

Dental Verbs often form present stem in -ζω. Those in -ίζω of more than two syllables have contracted fut. in -ιῶ (p. 48), *e.g.* νομίζω νομιῶ, but κτίζω κτίσω.

N.B. τ δ θ drop out before σ and κ.

κολάζ-ω 'I punish.' πείθ-ω 'I persuade.'

	Act.	*Pass.*	*Act.*	*Pass.*
Fut.	κολά-σω	κολασ-θήσομαι	πεί-σω	πεισ-θήσομαι
Aor.	ἐ-κόλα-σα	ἐ-κολάσ-θην	ἔ-πει-σα	ἐ-πείσ-θην
Perf.	κε-κόλα-κα	κε-κόλασ-μαι	πέ-πει-κα	πέ-πεισ-μαι

N.B. θαυμάζω 'I admire, wonder, am surprised' has fut. θαυμάσομαι (p. 30) and is often followed by an indirect question.

Verbs of teaching, concealing, begging, depriving may take two objects in the accusative in the Active (one in the Passive).

διδάσκω κρύπτω αἰτέω ἀπο-στερίσκω.

σώζ-ω I keep, save, preserve, *aor. pass.* ἐ-σώ-θην.
γράφ-ω I write, draw, *aor. pass.* ἐ-γράφ-ην.
πέμπ-ω I send, *perf. act.* πέ-πομφα, *perf. pass.* πέ-πεμ-μαι.
κρύπτ-ω I hide, *aor. pass.* ἐ-κρύβ-ην.

ἡ ἐλπίς ἐλπιδ- hope.
ἡ Ἑλλάς Ἑλλαδ- Greece.
ἡ πατρίς πατριδ- fatherland.
ἕνεκα for the sake of (*gen.*) (follows the noun).

τὸ ἅρμα ἁρματ- chariot.
τὸ χρῆμα χρηματ- thing.
τὰ χρήματα money.
ὑπέρ *prep. with acc.* over, beyond; *with gen.* on behalf of.

Ex. 38. 1. ὦ παῖ, τί δῆτα κέκρυφας ἐμὲ ταῦτα; τί οὐκ ἐπιστολὴν περὶ τούτων ἔγραψας;
2. οὐκ ἔφη μεμνῆσθαι ποῦ ἔκρυψε τὸν χρυσόν.
3. οἱ πιστοὶ στρατιῶται ὑπὲρ τῆς πατρίδος μάχονται, οὐ χρημάτων δὴ ἕνεκα ἀλλὰ βουλόμενοι σώζειν τούς τε παῖδας καὶ τὰς γυναῖκας, καὶ τὰς οἰκίας ὥστε μὴ βλάπτεσθαι αὐτοὺς ὑπὸ τῶν βαρβάρων.
4. ὁ στρατηγὸς λέγεται πεπομφέναι ἐκ τῆς Ἑλλάδος πολλοὺς τοξότας καὶ ἄλλους στρατιώτας· ὥστε μεγάλας δὴ ἐλπίδας νίκης ἔχομεν.
5. ὁ μὲν παῖς κολασθήσεται ὅτι ἔβλαψε τὸν ἵππον, τὴν δὲ κόρην, πεπεισμένην δὴ ὑπ' αὐτοῦ ταῦτα πράττειν, ἡμᾶς δεῖ οἰκτείρειν.
6. οἱ μὲν Ἰουδαῖοι ἐθαύμασαν οἷα ἐγράφη ὑπὸ τοῦ Πειλάτου περὶ τοῦ Ἰησοῦ· ὁ δὲ ἔφη, Ὃ γέγραφα, γέγραφα.
7. ἡ πίστις σου σέσωκέ σε.

Ex. 39. 1. Immediately the chariots were prepared and the army was drawn up for battle. For we were fighting, not persuaded (*perf. part.*) by the hope of money nor for the sake of gifts but on behalf of Greece, so as to prevent the barbarians from depriving us of our land.
2. Why, pray, did you conceal this from us?
3. I wondered how indeed the letter was concealed from the judge.
4. *Alexander tamed* a horse which others could not *overcome* (*aor.*).
5. He was saved by his own slave whom he himself once saved.
6. I hope the boy has not hurt the bird which he hid in the house.
7. They sent the good doctor and so he was saved when he was dying of a terrible disease.
8. I cannot remember his name.
9. We were much surprised that he did not send the money.
10. When the chariots were made, they were sent to Athens.

PERFECT PASSIVE OF VERBS WITH CONSONANT STEMS

Stem ταγ- Guttural		*Stem* κρυβ- Labial	
τέ-ταγ-μαι	τε-τάγ-μεθα	κέ-κρυμ-μαι	κε-κρύμ-μεθα
τέ-ταξαι	τέ-ταχ-θε	κέ-κρυψαι	κέ-κρυφ-θε
τέ-τακ-ται	τε-ταγμένοι εἰσί	κέ-κρυπ-ται	κε-κρυμ-μένοι εἰσί(ν)

Stem ψευδ- Dental		*Pluperfect* of ψευδ-	
ἔ-ψευσ-μαι	ἐ-ψεύσ-μεθα	ἐ-ψεύσ-μην	ἐ-ψεύσ-μεθα
ἔ-ψευ-σαι	ἔ-ψευσ-θε	ἔ-ψευ-σο	ἔ-ψευσ-θε
ἔ-ψευσ-ται	ἐ-ψευσ-μένοι εἰσί(ν)	ἔ-ψευσ-το	ἐ-ψευσ-μένοι ἦσαν

(1) These changes are made according to the table of consonants on p. 2. λ and ρ remain unchanged, ν sometimes changes to σ in the 1st person.

(2) The ἐ of ἐψεύσμην stands for both augment and reduplication.

(3) The participle follows the 1st person sing.; the infinitive follows the 2nd person plural, *e.g.* τε-ταγ-μένος, τε-τάχ-θαι.

REFLEXIVE PRONOUNS

αὐτός with personal pronouns forms reflexive pronouns.

Sing.	ἐγὼ αὐτός -ή	σὺ αὐτός -ή	αὐτός -ή
	ἐμαυτόν -ήν	σεαυτόν -ήν	ἑαυτόν -ήν -ό
	ἐμαυτοῦ -ῆς	σεαυτοῦ -ῆς	ἑαυτοῦ -ῆς -οῦ
	ἐμαυτῷ -ῇ	σεαυτῷ -ῇ	ἑαυτῷ -ῇ
Plur.	ἡμεῖς αὐτοί -αί	ὑμεῖς αὐτοί -αί	αὐτοί
	ἡμᾶς αὐτούς -άς	ὑμᾶς αὐτούς -άς	ἑαυτούς -άς -ά
	ἡμῶν αὐτῶν	ὑμῶν αὐτῶν	ἑαυτῶν
	ἡμῖν αὐτοῖς -αῖς	ὑμῖν αὐτοῖς -αῖς	ἑαυτοῖς -αῖς -οῖς

σεαυτόν and ἑαυτόν often contract into σαυτόν, αὑτόν.

ἕ, οὗ, οἷ are sometimes used instead of ἑαυτόν etc., and for plural in indirect speech σφᾶς, σφῶν, σφίσι(ν).

Special uses of αὐτός: (*a*) dative of manner, with a noun:

αὐτοῖς (τοῖς) ὅπλοις ἐσώθη 'He was rescued arms and all';

(*b*) with ordinal numerals:

στρατηγὸς ἐπέμφθη δέκατος αὐτός.
'He was sent as general with *nine* others.'

ψεύδ-ω I deceive; ψεύδομαι *middle* I lie (tell lies).
γενναῖ-ος -α -ον noble, brave. εἰ *conj.* if, εἰ μή unless.
ὁ ἡ θεός god, goddess. ὁ ἄνθρωπος man ('homo').
ὡς *conj.* as, when; *prep.* to (a person).

Note. Unlike Latin 'si,' εἰ can be used in an Indirect Question.

Ex. 40. 1. οὐκ ἔφη μεμνῆσθαι ὁ δοῦλος ποῦ ὁ χρυσὸς κέκρυπται. (Tense must be changed in English.) 2. ὁ κακὸς στρατιώτης ἔβλαψεν αὐτὸς ἑαυτὸν οὐ βουλόμενος μάχεσθαι τοῖς πολεμίοις. 3. ἐπέμψαμεν οὖν ἐπιστολὴν ὡς τὸν στρατηγὸν ἐν ᾗ ἐγέγραπτο τάδε· Εἰ μὴ αὐτίκα πέμψεις πολλοὺς ἄλλους, νικηθησόμεθα ὑπὸ τῶν Περσῶν. 4. πολλὰ πολλοὺς ἔψευσεν. 5. οὕτω δὴ τεταγμένοι ἐδύναντο ἢ ῥᾳδίως πορεύεσθαι ἢ εὐθὺς προσβάλλειν τοῖς πολεμίοις. 6. ποῦ ἐστὶν ὁ Ἐπισθένης; κῆρυξ πέπεμπται τρίτος αὐτός. 7. οὐδὲν γέγραπται ὑπὸ τῶν ποιητῶν γενναιότερον ἢ τὰ τοῦ Ὁμήρου ποιήματα. 8. ἤδη δὲ ἦν ἡμέρα καὶ ὁ στρατὸς ἐπορεύετο τεταγμένος ὡς ἐπὶ μάχην, οὕτω γὰρ ἐκεκέλευστο ὑπὸ τοῦ στρατηγοῦ. 9. χαλεπὸν μέν ἐστι ψεύδειν ἀνθρώπους ἀδύνατον δὲ Θεὸν αὐτὸν ψεῦσαι. (*Note the tense* ψεῦσαι = to deceive even once.) 10. τί δῆτα ἐψεύσω, ὦ παῖ; ἆρα μὴ ἤλπιζες κρύψειν ἐμὲ τὰ ὑπό σου πεπραγμένα; 11. οὗτοι δὴ τεταγμένοι εἰσὶν ἐν κακῷ χωρίῳ. 12. οἱ δ' οὖν ἡμέτεροι ἤρξαντο προσβάλλειν, ὁ μὲν γὰρ στρατὸς τῶν πολεμίων ἤδη ἐτέτακτο, τὰ δὲ ἅρματα παρεσκεύαστο, καὶ δὴ καὶ οἱ τοξόται ἐκπεπεμμένοι ἦσαν. 13. πῶς δῆτα σέσωσαι αὐτός, εἰ τὸ πλοῖον αὐτοῖς τοῖς ναύταις κεῖται ὑπὸ τῇ θαλάττῃ; αὐτὸς ἐμαυτὸν ἔσωσα, ἠπιστάμην γὰρ νεῖν, οἱ δὲ ἄλλοι κεκλεισμένοι ἐν τῷ πλοίῳ οὐκ ἐδύναντο σῴζειν ἑαυτούς.

Ex. 41. 1. He says that this letter has been written with hidden words so as to deceive the enemy. 2. I shall be surprised if he has been saved, for many in that village are dying of disease. 3. If you know how to swim, of course you can save yourself. 4. Many poems have been written (*Rule VI, p.* 12) by this poet. 5. The woman who pursued the *hen* into the village was deceived, for the egg (ᾠόν) had been cleverly concealed under the table in her own house. 6. If you wish to rule men well you ought to remember three-things (τριῶν): first, that you yourself are a man; second, that you should rule according to law; third, that you will not always rule. 7. We have sent two heralds concerning a treaty, for we wish to cease from war. 8. Some say that he transacted the affairs of the island well, but by others he is suspected to have deceived the islanders about the money by clever words. 9. The chariots had already been damaged and the phalanx itself had been thrown-into-confusion.

NOUNS WITH STEMS IN ν, ντ

In some of these the ς for nom. ending is dropped; in others the ν or the ντ is dropped and the vowel lengthened. In the dative plur. ν and ντ are always dropped and where two consonants fall out the vowel is lengthened, α becomes ᾱ but ο becomes ου and ε becomes ει.

	Sing.	*Plural*	*Sing.*	*Plural*
Nom.	ὁ μήν month	μῆν-ες	ὁ γέρων old man	γέροντ-ες
Acc.	μῆν-α	μῆν-ας	γέροντ-α	γέροντ-ας
Gen.	μην-ός	μην-ῶν	γέροντ-ος	γερόντ-ων
Dat.	μην-ί	μη-σί(ν)	γέροντ-ι	γέρουσι(ν)

Like γέρων are declined Participles with stems in -εντ, -οντ, -αντ, -υντ. The fem. follows the 1st decl. (θάλαττα), and the neuter *Nom. Voc. Acc.* end in -ν.

Participles in -ας:

Nom.	λύσας	λύσασα	λῦσαν	*aor. part. act.*
Acc.	λύσαντ-α	λύσασαν	λῦσαν	of λύω
Gen.	λύσαντ-ος	λυσάσης	λύσαντ-ος	
Dat.	λύσαντ-ι	λυσάσῃ	λύσαντ-ι	
Dat. pl.	λύσασι(ν)	λυσάσαις	λύσασι(ν)	

All *Participles* in -ων (*Present, Fut.* and *Strong Aor.*):

Nom.	ὤν	οὖσα	ὄν	'being'; *pres. part.* of the
Acc.	ὄντα	οὖσαν	ὄν	verb εἰμί ('I am').
Dat. pl.	οὖσι(ν)	οὔσαις	οὖσι(ν)	

Participles in -είς, λυθείς, λυθεῖσα, λυθέν. *Stem* λυθέντ-.

(1) It will be noticed that no meaning is given to the aor. part. λύσας for which neither 'loosing' nor 'having loosed' is really correct. The proper translation is to use a finite verb with a conjunction; or to prefix 'after' or 'when': κρύψας τὰ χρήματα ἀπέφευγεν 'he hid the money and fled'; πράξαντες ταῦτα ἐπορεύοντο 'when they had done this, they began to march.'

(2) The gen. absolute is often used in both act. and pass. τούτων λεχθέντων 'when this had been said'; κελεύοντος τοῦ στρατηγοῦ 'by order of the general'; πάντων παρεσκευασμένων 'when all was ready'; τούτων οὕτως ἐχόντων 'in these circumstances' ('quae cum ita sint').

Perf. Participles in -ώς:

N. V.	λελυκώς	λελυκυῖα	λελυκός
Acc.	λελυκότα	λελυκυῖαν	λελυκός
Dat. pl.	λελυκόσι(ν)	λελυκυίαις	λελυκόσι(ν).

πᾶς (also ἅπας and σύμπας) πᾶσα πᾶν, stem παντ- 'all, every, the whole.' The article should be used where it comes in the English: 'all the houses' πᾶσαι αἱ οἰκίαι but πᾶσα γυνή 'every woman'; 'everything' πάντα; 'all day' πᾶσαν τὴν ἡμέραν.

ὁ λιμήν λιμέν- harbour.
ὁ ἐλέφας ἐλέφαντ- elephant.
ὁ λέων λέοντ- lion.
ὁ χειμών χειμῶν- storm, winter.
χειμῶνος in winter time.
ὁρμάω I rush (start from *mid.*, ὁρμάομαι).
πλήν *adv.* but, *prep. with gen.* except.
ἅμα at the same time.
ἅμ' ἡμέρᾳ at daybreak.
*γε at least, at any rate.

Ex. 42. 1. τούτων δὲ οὕτως ἐχόντων ἐνομίσαμεν δεῖν ἡμᾶς παρασκευάσαντας τὰ ὅπλα πορεύεσθαι διὰ τοῦ πεδίου πρὸς τὴν κώμην.
2. ἐν τούτῳ πᾶσαι αἱ θύραι τῶν ἐν τῇ κώμῃ οἰκιῶν πλὴν τῆς ἐμῆς ἐκλείσθησαν.
3. ἐπεὶ δὲ πάντα παρεσκευάσθη ὅσα γε ὁ στρατηγὸς ἐκέλευσεν, οὐκέτι νυκτὸς οὔσης, ὥρμησαν ἐπὶ τοὺς πολεμίους.
4. εὖ γε ἐποίησας τὰ χρήματα κρύψας, ὦ γέρον.
5. ἅπας νεανίας νομίζει εἶναι σοφώτερος πάντων τῶν γερόντων.
6. πέμψας ταύτην τὴν ἐπιστολήν, ὁ Παυσανίας ἤλπιζε ποιήσεσθαι τὸν Ξέρξην φίλον ἑαυτῷ.
7. οἱ στρατηγοὶ ἔλεξαν ὅτι μετὰ τὴν ναυμαχίαν μεγάλου χειμῶνος ὄντος οὐκ ἐδύναντο σῴζειν τοὺς ἐν τῇ θαλάττῃ ναύτας· ἀλλ' ἔγωγε οὐκ ἐπίστευσα αὐτοῖς.
8. ἅμα ὁρμήσαντες ἐκ τοῦ λιμένος ἐδιώχθησαν ὑπὸ τῶν ἡμετέρων πλοίων.

Ex. 43. 1. In winter they hunt lions or elephants in that country. 2. We at any rate do not trust this man who, having deceived us many-times, wishes to accompany the army. 3. By the command of the judge, all the money was sent to Athens the same day. 4. After keeping quiet all that day, hidden in the house, the robbers rushed into the market-place and injured many old men, women and children. 5. On hearing this all of us except the judge rushed out of the market-place. 6. In the third month those (who were) in the prison were set free so that all the citizens wondered. 7. As there is a storm (*gen. absol.*) we shall perhaps be prevented from starting for two days and nights. 8. He saved himself, others he cannot save. 9. After having been shut up (*aor. part. pass.*) in prison for many months he sent a letter to (ὡς) his friends; and they (*p.* 18 (5)) ransomed him.

STRONG AORISTS

Many verbs form a strong aorist: act. in *-ον*, mid. *-όμην*, pass. *-ην*. These have the same personal endings, participle, and infin. as the corresponding tenses of λύω. When this happens the present stem is lengthened in some way. The meaning is the same whether a verb has a strong or a weak aorist. λείπω 'I leave,' aor. ἔ-λιπ-ον, infinitive λιπεῖν, partic. λιπών, λιποῦσα, λιπόν, stem λιπόντ-. The imperfect will be ἔλειπον. So deponent verbs: ὑπεσχόμην 'I promised,' ὑποσχέσθαι, ὑποσχόμενος. The indic. goes like an imperfect: ἔλιπ-ον, -ες, -ε, -ομεν, -ετε, -ον, and ὑπεσχόμην will follow ἐλυόμην.

Nouns in -ωρ, -ηρ are declined regularly except:

	Sing. father	*Plural*	*Sing.* man ('vir')	*Plural*
Nom.	πατήρ	πατέρες	ἀνήρ	ἄνδρες
Voc.	πάτερ	πατέρες	ἄνερ	ἄνδρες
Acc.	πατέρα	πατέρας	ἄνδρα	ἄνδρας
Gen.	πατρός	πατέρων	ἀνδρός	ἀνδρῶν
Dat.	πατρί	πατράσι(ν)	ἀνδρί	ἀνδράσι(ν)

Like πατήρ: μήτηρ 'mother'; θυγάτηρ 'daughter'; γαστήρ *fem.* 'belly.'

Note. ἔαρ *neut.* 'spring' (season), *gen.* ἦρος, *dat.* ἦρι.

USES OF PARTICIPLES

(1) Cause; often preceded by ἅτε (or ὡς with presumed reason). Neg. οὐ.

ὁ Κῦρος, ἅτε παῖς ὤν, ἥδετο τῇ στολῇ.

ὡς γὰρ τῶν πολεμίων πεφευγότων, ἄτακτοι ἐπορεύοντο.

(2) Concession; often preceded by καίπερ = though, or καὶ ταῦτα. Neg. οὐ.

ἑψόμεθα τούτῳ τῷ στρατηγῷ καίπερ οὐκ ὄντι Ἀθηναίῳ.

(3) Purpose; fut. partic. after verbs of motion. ὡς is prefixed after other verbs or to express a presumed purpose.

ὥρμησαν εἰς τὸ δεσμωτήριον λύσοντες τοὺς δεσμώτας.

παρεσκευάζοντο ὡς ἀπο-φευξόμενοι νυκτός.

(4) Condition (neg. μή): ταῦτα ποιήσαντες νικηθήσεσθε. When there is no negative translate 'by doing this.'

(5) Relative (neg. μή if indefinite) = he who..., people who....

οἱ τότε παρόντες ἐθαύμασαν τὰ πραττόμενα (or genitive)

οἱ διδάσκοντές εἰσι σοφώτεροι τῶν μανθανόντων.

Never use a pronoun with a participle, *e.g.* ἐκεῖνοι διδάσκοντες for 'those teaching.'

φεύγω flee	φεύξομαι	ἔ-φυγ-ον	πέ-φευγ-α
λείπω leave	λείψω	ἔ-λιπ-ον	λέ-λοιπ-α
λαμβάνω take	λήψομαι	ἔ-λαβ-ον	εἴληφα
εὑρίσκω find	εὑρήσω	εὗρ-ον	εὕρηκα
μανθάνω learn	μαθήσομαι	ἔ-μαθ-ον	μεμάθηκα
γίγνομαι happen, become	γενήσομαι	ἐ-γεν-όμην	γεγένημαι γέγονα
ἀφικνέομαι arrive	ἀφίξομαι	ἀφικόμην	ἀφῖγμαι
ἀπ-έ-θαν-ον I died, was killed		*infin.* ἀπο-θανεῖν	*partic.* ἀπο-θανών.
ἦλθον I came, went		*infin.* ἐλθεῖν	*partic.* ἐλθών.
εἶδον I saw		*infin.* ἰδεῖν	*partic.* ἰδών.

Ex. 44. 1. ἐπεὶ δ' ἀφίκοντο πρὸς τὴν κώμην μαθόντες τὰ ὑφ' ἡμῶν πραχθέντα ἐδίωκον πᾶσαν τὴν ἡμέραν.

2. ἐνταῦθα δὴ (thereupon) μάχης γενομένης πολλοὶ ἀπέθανον, οἱ δ' Ἀθηναῖοι ἔφυγον καταλιπόντες τὰ ὅπλα.

3. Γάϊος Καῖσαρ, Φαρνάκην τὸν Ποντικὸν νικήσας, ἔγραψε πρὸς τοὺς ἐν Ῥώμῃ πατέρας, Ἦλθον, εἶδον, ἐνίκησα.

4. καὶ νῦν πάρεσμεν ὡς ποιήσοντες ἃ ὑπεσχόμεθα.

5. ἐνταῦθα δεινὸν χρῆμα ἦν ἰδεῖν· οἱ μὲν γὰρ τῶν πολεμίων ἀπέθανον αὐτοὶ ὑφ' ἑαυτῶν, οἱ δὲ ἔρριψαν ἑαυτοὺς εἰς τὸν ποταμόν.

6. ὦ ἄνδρες Ἀθηναῖοι, ἔφη, νόμος ἐστὶν ἡμῖν τοὺς ταῦτα ποιήσαντας ἀποθανεῖν.

7. ἅμα φεύγοντες ἀπέρριπτον τὰ ὅπλα, καίπερ ἡμῶν οὐ διωκόντων.

8. ὁ δ' Ἀλκιβιάδης ἰδὼν τοὺς μὲν πολεμίους ἐν τῷ λιμένι ὄντας, τοὺς δὲ Ἀθηναίους οὐκ ἔχοντας σῖτον, ἔπειθεν αὐτοὺς ἀπὸ τοῦ κακοῦ χωρίου ἀπελθεῖν εἰς Σηστὸν ὥστε λιμένα τε ἔχειν καὶ σῖτον. οἱ δὲ στρατηγοὶ ἐκέλευσαν αὐτὸν ἀπελθεῖν· αὐτοὶ γὰρ νῦν στρατηγεῖν, οὐκ ἐκεῖνον.

9. εὕρηκα τὰ χρήματα ἃ ἔκρυψαν πορευσόμενοι.

Ex. 45. 1. When he found what he wanted he took the gold and fled. 2. So great a storm arose (ἐπι-γίγνομαι) that we were unable to find the road by-which we arrived. 3. Hearing what (had) happened he came forward and (παρελθών) spoke as follows. 4. Arriving on the third day we shall find everything well prepared. 5. He was persuaded by his father and mother to go to Athens. 6. What pray has happened? Our slaves have fled, taking (*aor. part.*) everything they could find. 7. Those who saw the lion immediately fled. 8. He punished (κολάζω) those who did not learn their work well.

CONTRACTED VERBS IN -αω

α + ο sound = ω. α + ε sound = ᾱ.
ι is written subscript. υ disappears.

τιμάομεν = τιμῶμεν, τιμάεις = τιμᾷς, τιμάουσι = τιμῶσι.
Contraction takes place only in the present stem.

Pres. Indic. Act. of τιμάω (τιμῶ) 'I honour'

τιμῶ (αω)	τιμῶμεν (αομεν)	*Infin.* τιμᾶν (no iota)
τιμᾷς (αεις)	τιμᾶτε (αετε)	*Part.* τιμῶν, τιμῶσα, τιμῶν,
τιμᾷ (αει)	τιμῶσι (αουσι(ν))	stem τιμῶντ-.

Exceptions: (1) χρῶμαι (χράομαι) 'I use, am friendly with'—contracts into η (for αε or αη) and forms its tenses with η contrary to *Rule III* (*p.* 6); fut. χρήσομαι, aor. ἐχρησάμην, perf. κέχρημαι.

(2) διψάω 'thirst,' πεινάω 'hunger,' ζάω 'I am alive,' also contract into η.

Ex. 46. Contract and give the meaning of the following: *Act.* ἐτίμαε, ἐτίμαον. *Pass.* ἐτιμάου, τιμάει.
Deponent χράεσθαι, χραόμενος, ἐχράετο.

NUMERALS. FIRST FOUR DECLINED

	one			two	three		four
N. V.	εἷς	μία	ἕν	δύο	τρεῖς	τρία	τέτταρ-ες -α
Acc.	ἕνα	μίαν	ἕν	δύο	τρεῖς	τρία	τέτταρ-ας -α
Gen.	ἑνός	μιᾶς	ἑνός	δυοῖν	τριῶν		τεττάρων
Dat.	ἑνί	μιᾷ	ἑνί	δυοῖν	τρισί(ν)		τέτταρσι(ν)

CARDINALS	ORDINALS	ADVERBS
πόσοι; how many?	πόστος; in what order?	ποσάκις; how often?
1. εἷς	πρῶτος	ἅπαξ
2. δύο	δεύτερος	δίς
3. τρεῖς	τρίτος	τρίς
4. τέτταρες	τέταρτος	τετράκις
5. πέντε	πέμπτος	πεντάκις
6. ἕξ	ἕκτος	ἑξάκις
7. ἑπτά	ἕβδομος	ἑπτάκις
8. ὀκτώ	ὄγδοος	ὀκτάκις
9. ἐννέα	ἔνατος	ἐνάκις
10. δέκα	δέκατος	δεκάκις
11. ἕνδεκα	ἑνδέκατος	ἑνδεκάκις
12. δώδεκα	δωδέκατος	δωδεκάκις

ὁ πούς ποδ- foot, *dat. pl.* ποσί(ν). πειράομαι *dep.* I try (*with infin.*)
ἡ χείρ χειρ- or χερ- hand, *dat. pl.* χερσί(ν).
πούς forms compounds δίπους τρίπους τετράπους etc.
οὐδείς (μηδείς) οὐδεμία οὐδέν = nobody, no one, *stem* οὐδεν- (μηδεν-).

Ex. 47. 1. πάντας τοὺς παῖδας δεῖ τιμᾶν τόν τε πατέρα καὶ τὴν μητέρα.

2. πόσα πλοῖα ἔχετε; ἔχομεν ἕνδεκα ἐν τῷ λιμένι.

3. ἆρ' ἐπίστασαι χρῆσθαι οἷς ἔχεις ὅπλοις;

4. τὰ δώδεκά ἐστι δὶς ἕξ, τρὶς τέτταρα, ἑξάκις δύο, τετράκις τρία.

5. εἰ βούλει τιμᾶσθαι αὐτός, δεῖ σε ἄλλους τιμᾶν.

6. ἐν τῇ θαλάττῃ ἔστι ζῷα δεινότατα οἷς ὄνομα ὀκτώποδες.

7. ζῶντα μὲν τοῦτον τὸν ποιητὴν ὕβριζον οἱ πολῖται, ἀποθανόντα δὲ νῦν πολὺ τιμῶσι.

8. 'Περὶ τοῦ τῆς Σφιγγὸς αἰνίγματος.'

ἡ μὲν Σφὶγξ πολὺ ἔβλαπτεν τὴν τῶν Θηβαίων χώραν διὰ τὸ μηδένα δύνασθαι λύειν τόδε τὸ αἴνιγμα· τί ζῷόν ἐστιν ἅμ' ἡμέρᾳ τετράπουν, ἐν μεσημβρίᾳ δίπουν, πρὸς νύκτα τρίπουν; τοῦτο δὲ τὸ αἴνιγμα ἐλύθη ὑπὸ τοῦ Οἰδίποδος· παῖς γάρ, ἔφη, ἐν ταῖς πρώταις ἡμέραις πειρᾶται χρῆσθαι τοῖς τε ποσὶ καὶ ταῖς χερσὶ βουλόμενος βαδίζειν· ἔπειτα δὲ ἀνὴρ γενόμενος τοῖς δύο ποσὶ μόνον· γέρων δὲ ὤν, χρῆται καὶ βάκτρῳ ὡς τρίτῳ ποδί. ἡ δὲ Σφὶγξ ἀκούσασα ταῦτα οὕτως ἀθύμως εἶχεν ὥστε ῥίψασα ἑαυτὴν εἰς τὴν θάλατταν ἀπέθανεν. ὁ δ' Οἰδίπους διὰ ταῦτα οὕτως ἐτιμήθη ὑπὸ τῶν Θηβαίων ὥστε ἐγένετο βασιλεύς.

Ex. 48. 1. Can you tell me how many feet a crocodile has?

2. He did not know how to use the arms he found.

3. For seven days our ships were prevented by a great storm from leaving the harbour but on the eighth they were able to start.

4. The Athenians took twelve ships and one of these crew and all.

5. For eleven furlongs they marched through the land without opposition (no one hindering, *gen. abs.*).

6. Those who are honoured by the citizens will receive many excellent gifts.

7. I think he is no longer alive because nothing has been announced about him for twelve days.

CONTRACTED VERBS IN -εω

ε + ε = ει, ε + ο = ου, ε + long vowel or diphthong is absorbed.

Pres. Indic. Act.	*Pres. Indic. Pass.*
φιλῶ (εω) love	φιλοῦμαι (εομαι) am loved
φιλεῖς (εεις)	φιλεῖ or ῇ (εει or εῃ)
φιλεῖ (εει)	φιλεῖται (εεται)
φιλοῦμεν (εομεν)	φιλούμεθα (εομεθα)
φιλεῖτε (εετε)	φιλεῖσθε (εεσθε)
φιλοῦσι(ν) (εουσι(ν))	φιλοῦνται (εονται)
Infin. φιλεῖν. *Partic.* φιλῶν.	*Infin.* φιλεῖσθαι. *Partic.* φιλούμενος.

Verbs in -εω with a stem of one syllable have only the εε = ει contraction.

πλέω 'sail,' πλεύσομαι, ἔπλευσα, πέπλευκα.

Imperfect of πλέω: ἔπλεον, ἔπλεις, ἔπλει, ἐπλέομεν, ἐπλεῖτε, ἔπλεον.

Pres. Infin. πλεῖν. Partic. πλέων, πλέουσα, πλέον.

So also πνέω 'breathe,' ῥέω 'flow' (aor. ἐρρύην), νέω 'swim.'

Note. καίω 'burn' (tr.), καύσω, ἔκαυσα etc.
κλαίω 'weep,' κλαύσομαι, ἔκλαυσα etc.

	Sing.		*Plural*		
	M F.	*N.*	*M.F.*	*N.*	who? what?
Nom.	τίς	τί	τίνες	τίνα	Indefinite also ἄττα
Acc.	τίνα	τί	τίνας	τίνα	
Gen.	τίνος or	τοῦ	τίνων		
Dat.	τίνι or	τῷ	τίσι(ν)		

τις corresponds to Latin 'quis.'

The interrogative τίς; has throughout the acute accent on the ι, the indefinite never. Both may be used alone or as adjectives. The indefinite τις must always follow some other word:

ποῖός τις; of what sort? πᾶς τις, every one.
ἄλλοι τινές, some others. εἴ τις, if anyone.
φιλόσοφός τις, a (certain) philosopher.

ὁ ἡ κύων κυν- dog, *voc.* κύον, *dat. pl.* κυσί(ν).
τὸ πῦρ πυρ- fire, *plural 2nd decl. neuter*, watch-fires.
τὸ ὕδωρ ὑδατ- water, *dat. pl.* ὕδασι(ν).
ἕκαστος -η -ον each.
κινέω move (*trans.*).
ὠφελέω help.
φοβέω terrify.
φοβέομαι fear.
μᾶλλον more, rather.
ἑκάτερος -α -ον each of two.
πολιορκέω besiege.
δέ-ομαι beg (*acc. and infin.*), need (*gen.*).
δοκέω seem, think, (*impersonal or with dative*) seem good to.
κατοικέω inhabit.

Ex. 49. 1. κατὰ τὸν Ἡράκλειτον πάντα ῥεῖ ἀεὶ καὶ κινεῖται, καὶ οὐδὲν μένει τὸ αὐτό.

2. πολλὰ καὶ μεγάλα ἀδικεῖς ἡμᾶς ὃς ταῦτα ποιεῖς.

3. οὐ μόνον ὁ ποιῶν τι ἀλλὰ καὶ ὁ μὴ ποιῶν τι πολλάκις ἀδικεῖ.

4. ἐν δὲ τούτῳ οἱ στρατιῶται μᾶλλον πολιορκούμενοι ἢ πολιορκοῦντες ἀθύμως εἶχον.

5. οἱ κύνες ἅπαξ δὴ καυθέντες λέγονται φοβεῖσθαι τὸ πῦρ.

6. τίνες ἐστέ; καὶ ἀπὸ τίνος χώρας; Ἀθηναῖοί ἐσμεν, καὶ εἴ γέ σοι δοκεῖ, βουλόμεθα κατοικεῖν τὴν κώμην εὑρήσοντες σῖτον καὶ ὕδωρ.

7. εἴ τις, πατὴρ ὤν, μὴ φιλεῖ τοὺς παῖδας, τοῦτον δὴ δοκοῦμεν χαλεπὸν καὶ αἰσχρόν.

8. ἐπεὶ ἀδύνατον ἦν δι-εκ-πλεῖν ἐκ τοῦ λιμένος ἐδόκει ἡμῖν κατακαῦσαι τὰ πλοῖα καὶ ἐκφυγεῖν ἀνὰ τὴν χώραν.

9. οὐκ ἐπειρώμεθα διαβαίνειν τὸν ποταμόν, πολὺς γὰρ ἔρρει.

10. οἱ μὲν δὴ ταῦτα ἐποίουν· οἱ δ' ἐν τῷ χωρίῳ, καὶ γὰρ πολὺν χρόνον ἐπολιορκοῦντο καὶ ὁ σῖτος ἐπ-έλιπεν, κήρυκα ἔπεμψαν περὶ σπονδῶν.

11. ἐν τούτῳ τῷ χρόνῳ ἤδη δύο μηνῶν ὄντων ἀφικνοῦνται κήρυκες παρὰ Θίβρωνος οἳ λέγουσιν ὅτι Λακεδαιμονίοις δοκεῖ στρατεύειν ἐπὶ τοὺς Πέρσας· ὁ γὰρ Θίβρων ἐκπεπλευκὼς ὡς πολεμήσων δεῖται ταύτης τῆς στρατιᾶς καὶ δαρεικὸς στρατιώτῃ ἑκάστῳ ἔσται μισθὸς (pay) τοῦ μηνὸς τοῖς δὲ στρατηγοῖς τετρα-μοιρία.

Ex. 50. 1. My boy, you ought to love and honour your father and mother since they love you and have always helped you. 2. Those who have much do not need money but the rich ought to help those who have nothing. 3. He rushed into the road saying, "My house is on fire and I am unable to free my horse. I entreat you all to bring water immediately." 4. While we were sailing to Egypt the ship was burnt and I alone have escaped. 5. He says he cannot remember what sort of dogs you ordered him to send. 6. Some one dared to throw himself into the sea and save the terrified boy. 7. "What did you do in the great war?" said a (certain) boy to his father.

CONTRACTED VERBS IN -οω

Rule. ο + ο or ε = ου, ο + ω or η = ω. Any combination with ι = οι.

δηλόω 'make plain'

Pres. Indic. Act.	*Pres. Indic. Pass.*
δηλῶ (οω)	δηλοῦμαι (οομαι)
δηλοῖς (οεις)	δηλοῖ (οει or οῃ)
δηλοῖ (οει)	δηλοῦται (οεται)
δηλοῦμεν (οομεν)	δηλούμεθα (οομεθα)
δηλοῦτε (οετε)	δηλοῦσθε (οεσθε)
δηλοῦσι(ν) (οουσι(ν))	δηλοῦνται (οονται)
Infin. δηλοῦν. *Partic.* δηλῶν.	*Infin.* δηλοῦσθαι. *Partic.* δηλούμενος.

SECOND DECLENSION NOUNS CONTRACTED

	mind, brains, sense.				bone.			
Nom.	ὁ νοῦς	(οος)	νοῖ	(οοι)	τὸ ὀστοῦν	(εον)	ὀστᾶ	(εα)
Voc.	νοῦ	(οε)	νοῖ	(οοι)	ὀστοῦν	(εον)	ὀστᾶ	(εα)
Acc.	νοῦν	(οον)	νοῦς	(οους)	ὀστοῦν	(εον)	ὀστᾶ	(εα)
Gen.	νοῦ	(οου)	νῶν	(οων)	ὀστοῦ	(εου)	ὀστῶν	(εων)
Dat.	νῷ	(οῳ)	νοῖς	(οοις)	ὀστῷ	(εῳ)	ὀστοῖς	(εοις)

Similarly adjectives in -εος, -οος contract.

χρύσεος χρυσέα χρύσεον golden.
χρυσοῦς χρυσῆ χρυσοῦν,

but ἀργύρεος ἀργυρέα ἀργύρεον of silver.
ἀργυροῦς ἀργυρᾶ ἀργυροῦν (according to *Rule III, p.* 6).

Price and value are expressed by the genitive after verbs and adjectives of buying, valuing and selling. The following verbs are irregular:

πωλέω 'I sell,' ἀποδώσομαι, ἀπεδόμην, πέπρακα, aor. pass. ἐπράθην.

ὠνέομαι 'I buy,' ὠνήσομαι, ἐπριάμην, ἐώνημαι.

The imperfect of ὠνέομαι is ἐωνούμην. ἀγοράζω is used of buying or selling in a market. τιμάω = I value. ἄξιος -α -ον = worth.

UNIVERSAL RELATIVE: ὅστις 'whosoever'

N.	ὅστις	ἥτις	ὅτι (ὅ,τι)	οἵτινες	αἵτινες	ἅτινα or ἅττα
A.	ὅντινα	ἥντινα	ὅτι (ὅ,τι)	οὕστινας	ἅστινας	ἅτινα or ἅττα
G.	ὅτου	ἧστινος	ὅτου	ὅτων	ὧντινων	ὅτων
D.	ὅτῳ	ᾗτινι	ὅτῳ	ὅτοις	αἷστισι(ν)	ὅτοις

This is also used in Indirect Questions for τίς;

αὖθις (*adv.*) again.
ἀπ-αιτέω I demand.
ἡ ζημία penalty.
ἡ φωνή sound, speech.
ἡ σιωπή silence.
ἡ δραχμή drachma (9½*d.*)
ἡ μνᾶ (*gen.* μνᾶς) mina (£4).
τὸ τάλαντον talent (£240).
ἡ βίβλος, τὸ βιβλίον roll, book.
πάλιν (*adv.*) back.
τίμιος -α -ον valuable.
ὁ Θρᾷξ Θρᾳκ- Thracian.
ὁ τύραννος absolute ruler, tyrant.
πόσου; for how much?
τότε (*adv.*) then.
ἡ ἀξία, ἡ τιμή price, value.

Ex. 51. 1. τίς ἔφη τόδε; ἡ μὲν φωνή ἐστιν ἀργυρᾶ, ἡ δὲ σιωπὴ χρυσῆ.

2. οἱ Θρᾷκες ἐωνοῦντο τὰς γυναῖκας παρὰ τῶν πατέρων χρημάτων πολλῶν.

3. πόσου πωλεῖς τάδε; μιᾶς δραχμῆς ἕκαστον.

4. βούλομαι ἀποδόσθαι τὴν οἰκίαν δύο ταλάντων, ἐπριάμην γὰρ αὐτὴν πολλοῦ.

5. Ταντάλου μετὰ θάνατον ἡ ζημία ἦν ἀεὶ διψῆν καὶ πεινῆν (p. 46).

6. *The Sibyl's prophecies.*

τυράννου τινός ποτε ἄρχοντος τῶν Ῥωμαίων, λέγεται προφῆτις τις, ὀνόματι Σίβυλλα, φέρουσα ἐννέα βίβλους, πειράσασθαι αὐτὰς πωλεῖν τριακοσίων μνῶν. ὁ δὲ τύραννος οὐκ ἤθελε ὠνεῖσθαι τοσαύτης ἀξίας. ἡ δ' οὖν γυνὴ τότε μὲν ἀπῆλθεν· τῇ δὲ ὑστεραίᾳ πάλιν ἐλθοῦσα ἀπῄτει τὴν αὐτὴν τιμὴν καίπερ ἓξ μόνον βίβλους προσφέρουσα. ὁ δὲ τύραννος αὖθις δὴ οὐκ ἔφη ὠνήσεσθαι αὐτάς. τῇ δὲ τρίτῃ ἡμέρᾳ ἡ Σίβυλλα, καίπερ τριῶν μόνον τῶν βίβλων λοιπῶν ὄντων, τὰς τριακοσίας μνᾶς ὅμως (still) ἀπῄτησεν.

Ex. 52. 1. But when the tyrant, naturally (p. 36), wondered, the Sibyl said that she had burnt the six books but that the rest, on that account, were much more valuable. Thereupon the tyrant, obeying her, bought them for 300 minae. And she, bidding him keep them and (*partic.*) hide them well, departed. And always after that the Romans, consulting (χράομαι) these books, used to learn what they ought to do, so that often they were able to overcome the attacks of their enemies (*pres. partic.*).

2. They shall have (εἰμί *with dat.*) ten drachmas each.

3. You are enslaving men who fought well in the war against (πρός *with acc.*) the Persians.

4. Do you not reckon (it) of great importance to try to be honoured by all the citizens?

6. At what price did you buy that horse?

THIRD DECLENSION NEUTER CONTRACTED NOUNS IN -ος

Stem in -ες; the σ is dropped except in the nom. sing.

N. V. A.	γένος race, kind	γένη (γενεσα)
Gen.	γένους (γενεσος)	γενῶν (γενεσων)
Dat.	γένει (γενεσι)	γένεσι (γενεσσι)

ADJECTIVES FORMED FROM NOUNS

	M. F.	*N.*		*M. F.*	*N.*
Nom.	εὐγενής	εὐγενές	well-born	εὐγενεῖς	εὐγενῆ
Voc.	εὐγενές	εὐγενές		εὐγενεῖς	εὐγενῆ
Acc.	εὐγενῆ	εὐγενές		εὐγενεῖς	εὐγενῆ
Gen.	εὐγενοῦς			εὐγενῶν	
Dat.	εὐγενεῖ			εὐγενέσι	

Those in -ης not formed from these neuter nouns have the usual declension, *e.g.* πένης, πένητος 'poor (man).'

Many proper names were formed like εὐγενής and declined accordingly: Διογένης, Ἀριστοτέλης, Σωκράτης.

The following verbs in -έω form their principal tenses with -ε not -η. ἐπ-αινέω 'praise,' ἐπαινέσω, ἐπῄνεσα etc.

τελέω 'finish,' τελέσω, ἐτέλεσα, aor. pass. ἐτελέσθην.

καλέω 'call,' καλέσω, ἐκάλεσα, κέκληκα etc.

The futures of these in Attic Greek contract into -ῶ.

There are some in -άω which form tenses with short ᾰ: γελάω 'I laugh,' γελάσομαι, ἐγέλασα. καταγελάω 'I laugh at' like many verbs compounded with κατά takes genitive.

IRREGULAR COMPARISON IN -ίων, -ιστος

μέγας 'great,' μείζων, μέγιστος.

Nouns and adjectives in -ων, stem -ον, are declined regularly (dat. pl. -οσι, *e.g.* ἡ χιών 'snow,' gen. χιόνος and the corresponding adjectives similarly; neut. sing. -ον, neut. pl. -ονα, *e.g.* σώφρων 'prudent,' pl. σώφρονες, neut. σώφρονα).

Comparatives in -ων, however, contract: μείζονα into μείζω; μείζονες and μείζονας into μείζους.

DIMINUTIVES are formed chiefly in -ιον, -άριον, -ίδιον, -ίσκος. παῖς, παιδ-, παιδίον, παιδάριον, παιδίσκος, 'little boy'; παιδίσκη 'little girl'; ἀνθρωπίσκος 'mannikin.'

κύων, κυν-, κυνίδιον 'little dog'; (Ἑρμῆς) Ἑρμίδιον 'small statue of Hermes.'

τὸ θέρος heat, summer.
τὸ ἔτος year.
τὸ ὄρος mountain.
τὸ τέλος end.
τὸ τεῖχος wall.
ἡ τριήρης trireme.
ὅμως *adv.* still, nevertheless.
ὅμοιος -α -ον like (*with dat.*).
δουλόω I enslave.
ἀξιόω deem worthy, claim.
τειχίζω fortify.
ἀληθής -ές *adj.* true.
ἀσφαλής -ές safe.
εὐτυχής -ές fortunate.
ὁ ἀδελφός brother.
μήν *adv.* assuredly

Ex. 53. 1. τῷ δὲ αὐτῷ θέρει οἱ Πέρσαι ἐδούλουν πάσας τὰς κώμας ὅσαι γε οὐ τετειχισμέναι ἦσαν, τὴν δὲ ἡμετέραν, μείζω τείχη ἔχουσαν, οὐκ ἐδύναντο δουλοῦν.

2. πολλὰ μὲν ἔτη ἀσφαλῶς καὶ εὐτυχῶς κατῳκοῦμεν ἐν ταύτῃ τῇ νήσῳ, τέλος δὲ οἱ πολέμιοι μεγίστας τριήρεις ἔχοντες ἐφ' ἡμᾶς ἐπιπλεύσαντες ἔλαβον τὴν νῆσον οὖσαν ἀτείχιστον (*Rule VII, p.* 14).

3. ὅστις ἀεὶ τὰ ἀληθῆ λέγει τούτῳ πάντες πιστεύουσιν.

4. ἆρ' οὐκ ἀξιοῖς τὸν Σωκρατῆ εἶναι σοφώτατον; ἀληθῆ λέγεις, ὅμως δὲ ἀπέθανεν ὑπὸ τῶν Ἀθηναίων.

5. πολὺν μὲν χρόνον ἐφύλαττον τὰ τείχη· τέλος δὲ ἐπεὶ οὐκ ἀσφαλὲς οὐκέτι ἐδόκει αὐτοῖς, τῶν πολεμίων πολλῶν ὄντων, ἔφυγον εἰς τὰ ὄρη νυκτός.

6. ἦν ποτε κώμη τις, Κύμη ὀνόματι, ἧς οἱ ἄλλοι Ἕλληνες κατεγέλων τῶν ἐνοικούντων ὡς οὐδένα νοῦν ἐχόντων ἢ πάνυ γε μικρότατον· εἷς δὲ τῶν Κυμαίων ἰδών ποτε δύο ἀδελφούς, τῶν παρόντων τινῶν θαυμαζόντων τὴν ὁμοιότητα αὐτῶν, Οὐχ οὕτως ὅμοιος, ἔφη, οὗτός ἐστιν ἐκείνῳ, ὡς ἐκεῖνος τούτῳ.

Ex. 54. 1. Those-men therefore will never be fortunate since they do not dare to speak the truth. 2. Those who live in the mountains are poor, but still fortunate; for the enemy cannot enslave them. 3. What man is so safe that he (ὅστις) can never be injured by the evil deeds of other men? 4. He says that his name is Anaxagoras and that he is (an) Athenian by race (*acc. or dat.*). 5. This silver coin (νόμισμα) is very like those which we found at Cume lying among the bones of men. 6. For three years we were enslaved but at last we were freed by a general sent from Athens with twelve triremes. 7. Why do you not make plain to them that you claim to keep the money which you found? 8. These mountains are much greater than those of Greece.

VERBS WITH STEMS ENDING IN λ, ρ, μ, ν

The Fut. Act. was formed in -έσω and the Fut. Middle in -έσομαι, but the σ dropped and contraction followed as in φιλέω.

σημαίνω 'I signal,' Stem σημαν-, Fut. σημαν-έσω = σημανῶ.

φαίνομαι 'I appear,' φανέσομαι = φανοῦμαι, Aorist ἐφάνην.

N.B. In Attic Greek verbs in -ίζω have also this contracted future: κομίζω, fut. κομιῶ; ψηφίζομαι 'vote for,' ψηφιοῦμαι.

The Aorist Act. and Mid. were formed with the loss of σ and the lengthening of vowel: ἐσημαν-σα became ἐσήμηνα; μένω Aor. ἔμεινα.

Here *Rule III* applies; ὑγιαίνω, Aor. ὑγίᾱνα. The remaining tenses when required were formed regularly, *e.g.* σεσήμαγκα, but ε was sometimes changed before λ, ρ, into ᾰ and the Aor. Pass. often is formed strong with -ην not -θην. δια-φθείρω 'I destroy,' -φθερῶ, -έφθειρα, -έφθαρκα, -έφθαρμαι, -εφθάρην.

The following very common verbs should be learnt:

ἀπο-κτείνω 'kill,' -κτενῶ, -έκτεινα, -έκτονα. Passive supplied by ἀπο-θνήσκω 'am killed,' -θανοῦμαι, -έθανον, τέθνηκα 'I am dead.'

μένω 'remain,' μενῶ, ἔμεινα.

ἀγγέλλω 'announce,' ἀγγελῶ, ἤγγειλα, ἤγγελκα etc.

κρίνω 'judge, decide,' κρινῶ, ἔκρινα, κέκρικα etc.

ἀπο-κρίνομαι 'reply,' -κρινοῦμαι, -εκρινάμην, -κέκριμαι.

βάλλω 'throw, pelt,' βαλῶ, ἔβαλον, βέβληκα etc.

ἀπο-στέλλω 'send off,' -στελῶ, -έστειλα, -έσταλκα etc.

ὁ θάνατος death.	ὁ βίος life.
ἀ-θάνατος -ον deathless.	θνητός -ή -όν mortal.
ἡ ψυχή life, soul.	τὸ σῶμα body.
ἔρχομαι I go, *aor.* ἦλθον.	εἰσβάλλω εἰς invade.

Ex. 55. 1. ὁ δὲ ἀπεκρίνατο· Ἆρ' οὐ νομίζεις τὴν μὲν ψυχὴν τῶν ἀνθρώπων εἶναι ἀθάνατον τὸ δὲ σῶμα θνητόν· τοῦ δὲ σώματος διαφθαρέντος τὴν ψυχὴν οὐκ ἀποθνήσκειν ἀλλ' ἀπ-έρχεσθαι εἰς ἄλλον τινὰ βίον; 2. μετὰ δὲ ταῦτα ἤγγειλαν οἱ κήρυκες ὅτι οἱ πολέμιοι εἰσβαλόντες εἰς τὴν χώραν ἡμῶν διαφθείρουσι πάντα. 3. τούτων οὖν οὕτως ἐχόντων ἔμεινεν ἐν τούτῳ τῷ χωρίῳ τρεῖς ἡμέρας· ὕστερον δὲ ἐπορεύθη διὰ τοῦ πεδίου πρὸς τὰ ὄρη οὐδενὸς κωλύοντος. 4. ἐπειδὴ πάντα παρεσκευάσθη ὁ στρατηγὸς ἀπέστειλε δέκα τριήρεις πρὸς τὴν νῆσον ᾗ οἱ Πέρσαι προσέβαλλον.

εὐκλεής -ές famous.
ἡ τέχνη skill, art.
τὸ πινάκιον tablet, canvas.
τὸ γραφεῖον pencil, brush.
τὸ κιβώτιον box.
θάπτω bury, *aor. pass.* ἐτάφην.
κατα-κρίνω condemn.
λευκός -ή -όν white.
φοιτάω visit, attend lectures.
ἡ ἐπιμέλεια trouble, pains.
καινός -ή -όν new, fresh.
τὸ χρῶμα colour, paint.
ἐμφαίνω indicate, emphasise.
ἡ ψῆφος pebble.
μέλας μέλαινα μέλαν black.
ὁ ξένος stranger, guest, host.

With Brains, sir!

Ex. 56. 1. ἦν ποτὲ ζωγράφος (painter) τις εὐκλεής, Ἀντισθένης ὀνόματι. ἐπειδὴ δὲ νεανίσκος τις βουλόμενος φοιτᾶν παρ᾽ αὐτὸν καὶ ἐλπίσας μαθεῖν τὴν ζωγραφικὴν τέχνην ῥᾳδίως καὶ ἄνευ πολλῆς ἐπιμελείας, ἠρώτησε τίνων δεῖ ὥστε γενέσθαι ζωγράφος εὐκλεής, ὁ Ἀντισθένης ἀπεκρίνατο, Δεῖ πινακίου καινοῦ καὶ γραφείου καινοῦ καὶ χρωμάτων κιβωτίου καινοῦ, τὸν νοῦν ἐμφαίνων.

2. ἐψηφίσαντο δὲ οἱ κριταὶ αὐτὸν ἀποθανεῖν· ἀποθανόντα δὲ ἔθαψαν· οὐδεὶς δὲ τάφος αὐτοῦ οὐδέποτε ἐφάνη. οἱ δὲ ψηφιζόμενοι ἔβαλλον εἰς κιβώτιόν τι, οἱ μὲν κατακρίνοντες ψήφους μελαίνας, οἱ δὲ ἀπολύοντες μὴ ἀδικεῖν ψήφους λευκάς.

Ex. 57. 1. Thereupon Demosthenes coming forward (παρελθών) spoke as follows: "I for my part (ἔγωγε) do not consider that it is safe to vote thus, fellow-citizens; for by doing (*aor. part.*) this you will destroy not only yourselves but your wives and children, since the barbarians will kill all those who try to prevent them from entering (εἰσελθεῖν) the village."

2. In Cyme when a certain famous man was being buried, *a* stranger coming forward asked: "Who is the dead man (τεθνηκώς)?" Thereupon one of the citizens replied, "That man who lies upon the bier (ἡ κλίνη)."

3. He is having his sons taught the art of painting.

4. Since the bodies of men are mortal we shall all die but our souls, being immortal, cannot die.

5. The herald announced that the ships were all destroyed by a great storm.

NOUNS AND ADJECTIVES WITH SOFT VOWEL STEMS IN -ι, -υ

	Sing. city, state.	*Plur.*	*Sing.* elder.	*Plur.* ambassadors.
Nom.	ἡ πόλις	πόλεις	ὁ πρέσβυς	πρέσβεις
Voc.	πόλι	πόλεις	πρέσβυ	πρέσβεις
Acc.	πόλιν	πόλεις	πρέσβυν	πρέσβεις
Gen.	πόλεως	πόλεων	(πρέσβεως)	πρέσβεων
Dat.	πόλει	πόλεσι(ν)	(πρέσβει)	πρέσβεσι(ν)

Nouns ending in -σις formed from verbs, all feminine, are declined like πόλις, but only ὁ πέλεκυς 'axe,' ἔγχελυς 'eel,' πῆχυς 'fore-arm,' 'cubit,' follow πρέσβυς. The rest in -υς have acc. -υν but otherwise follow consonant stems.

ADJECTIVES

	sweet					
Nom.	ἡδύς	-εῖα	-ύ	ἡδεῖς	ἡδεῖαι	ἡδέα
Acc.	ἡδύν	-εῖαν	-ύ	ἡδεῖς	ἡδείας	ἡδέα
Gen.	ἡδέος	-είας	-έος	ἡδέων	ἡδειῶν	ἡδέων
Dat.	ἡδεῖ	-είᾳ	-εῖ	ἡδέσι(ν)	ἡδείαις	ἡδέσι(ν)

IMPERSONAL VERBS

ἔξεστί μοι 'it is permitted'
συμφέρει μοι 'it is expedient'
μέλει μοι τούτου 'this concerns me'
λυσιτελεῖ μοι 'it is profitable'

} all can have an infin. for subject.

Where a genitive absolute would be used with an ordinary verb, the neuter participle of the Impersonal verb stands alone:

ἐξόν μοι τοῦτο ποιῆσαι, ὅμως οὐ ποιήσω 'It being permitted (= though I might do this), still I will not.'

MEASUREMENT

(*a*) Distance is expressed by the acc., but (*b*) dimensions can be in the genitive—width, height etc. being in the acc., (*c*) before a comparative however a dative is required.

(*a*) *ἐντεῦθεν* (thence) *ἐξελαύνει σταθμοὺς τρεῖς, παρασάγγας εἴκοσιν εἰς Ἰκόνιον τῆς Φρυγίας ἐσχάτην πόλιν.*

(*b*) *διὰ μέσης δὲ τῆς πόλεως ῥεῖ ποταμὸς Κύδνος ὄνομα, τὸ εὖρος δύο πλέθρων* or *εὐρὺς δύο πλέθρα.*

(*c*) *πολλῷ χρησιμώτερος. κεφαλῇ μείζων* = taller by a head.

πίπτω fall, πεσοῦμαι, ἔπεσον, πέπτωκα.

ἡ τάξις arrangement, rank.
ἡ ἀναχώρησις retreat (ἀνα-χωρέω).
ὁ ἐνιαυτός year.
τὸ ξύλον wood = timber.
ἡ δύναμις power.
ὁ λίθος stone.
ὁ ἡ ἰχθύς fish.
τῇ ὑστεραίᾳ on the next day.

Ex. 58. 1. Ἀγησίλαος ἐρωτηθεὶς διὰ τί ἀτείχιστός ἐστιν ἡ Σπάρτη, Οὐ λίθοις δεῖ καὶ ξύλοις τετειχίσθαι τὰς πόλεις, ἔφη, ταῖς δὲ τῶν ἐνοικούντων ἀρεταῖς.

2. οὐκ ἀεὶ συμφέρει ποιεῖν ὃ σοι ἔξεστιν.

3. τοῦ δὲ ἐπι-γιγνομένου θέρους ἡ νόσος τὸ δεύτερον ἐπ-έπεσε τῇ πόλει· παρέμεινε δὲ κατ' ἐνιαυτόν, ὥστε πολὺ ἐκάκωσε τὴν τῶν Ἀθηναίων δύναμιν. πολλοὶ γὰρ ἀπέθανον ἐκ τῶν τάξεων τοῦ στρατοῦ· καὶ ἀριθμὸς τῶν πολιτῶν διεφθάρη ἀν-εξ-εύρε-τος.

4. ἐνόμιζον οἱ Ἕλληνες ἑκάστῳ τῶν πολιτῶν μέλειν ὡς ἰσχυροτάτην ποιεῖν τὴν πόλιν.

5. πολλοῖς μὲν θέρους δοκεῖ τὸ τοὺς ἰχθύας λαμβάνειν ἄλλοις δὲ ἡδύ ἐστι πλεῖν ἢ νεῖν ἐν τῇ θαλάττῃ.

6. μετὰ δὲ τὴν τῶν Περσῶν ἀναχώρησιν ἐδόκει τοῖς Ἀθηναίοις συμφέρειν ἑαυτοῖς περι-τειχίζειν τὴν πόλιν· οἱ δὲ Λακεδαιμόνιοι ἐπειρῶντο κωλύειν αὐτοὺς οἰκοδομεῖσθαι τὰ τείχη. ἠξίουν τε αὐτοὺς μὴ τειχίζειν πέμψαντες ἀγγέλους· οἱ δὲ ἀπεκρίναντο ὅτι πέμψουσι καὶ αὐτοὶ πρέσβεις περὶ ὧν λέγουσιν.

7. οὐ λυσιτελεῖ μοι πωλεῖν τοὺς ἰχθύας ταύτης τῆς τιμῆς.

Ex. 59. 1. Each of the Greeks thought (it) a noble (thing) to fight and die for the city. 2. Since the ranks were broken, it seemed (good) to the general to retreat to a stronger place. 3. This city was two furlongs distant from the sea. 4. Thus therefore they departed each to his own city. 5. In each rank there were twelve men armed with great spears (λόγχη). 6. On the next day they marched 20 furlongs and came to *a* river which (it) was impossible to cross without boats, because it was twelve feet deep (βαθύς). 7. Why do you try to make a treaty, it being possible (παρόν) to invade the country without opposition? 8. I do not consider that all this concerns either me or you. 9. It had not been announced that the treaty had been broken; therefore we were not guarding the walls of our city. 10. Each city was ordered to send troops according to (its) power. 11. It is pleasant in summer to swim in the river. 12. These barbarians are said to worship (σέβομαι) gods (made) of wood and stone.

THE IMPERATIVE MOOD

		Pres. Active	Aor.	Pres. Mid. and Passive
Sing.	2.	λῦε	λῦσον	λύου (λυ-εσο)
	3.	λυέτω	λυσάτω	λυέσθω
Plur.	2.	λύετε	λύσατε	λύεσθε
	3.	λυόντων	λυσάντων	λυέσθων

		Aor. Mid.	Aor. Pass.
Sing.	2.	λῦσαι	λύθητι (for λύθη-θι)
	3.	λυσάσθω	λυθήτω
Plur.	2.	λύσασθε	λύθητε
	3.	λυσάσθων	λυθέντων

(*a*) The perfect has an Imperative when required, *e.g.* for tenses with a perfect form and a present meaning. For the perfect active the endings will be the same as the present, added to the stem λελυκ-. For the perfect middle and passive the imperative, for instance, of *μέμνημαι* 'I remember' is *μέμνη-σο, μεμνή-σθω, μέμνη-σθε, μεμνή-σθων.*

(*b*) Imperative of *εἰμί* ('sum') = *ἴσθι, ἔστω,* plur. *ἔστε, ὄντων.*

(*c*) The aorist, as usual, denotes a single act without reference to time (*ἀ-όριστος* = indefinite); the present emphasises the continuance of the verbal action, *e.g.*

κροῦσον ἐκείνην τὴν μυῖαν = swat that fly!
ἀεὶ δίωκε τὴν ἀρετήν = always pursue virtue.

Contracted verbs only contract the present-stem tenses. The present imperative of *τιμάω* is therefore *τίμα, τιμάτω, τιμᾶτε, τιμώντων,* but the imperative aorist follows λῦσον (see above).

N.B. The genitive and dative plural of the participle are easily mistaken for other parts of the verb, *i.e.* λυόντων 3rd plur. imperat. pres. or gen. plur. of the participle; λύουσι dat. plur. of the pres. partic. or 3rd plur. pres. indic. Note also that the 2nd plur. pres. indic. and pres. imperat. are spelt the same—λύετε.

SPECIAL CAUTION

The 3rd person of the Imperative is used for 'let him, let them...,' not the Subjunctive as in Latin.

-έστερος -έστατος.

Adjectives with a stem in *-ον* and those with a stem in -ες are compared: *σώφρων* 'prudent' *σωφρονέστερος, ἀληθής* 'true' *ἀληθέστατος. ὅτι* and *ὡς* are often prefixed to superlatives (as 'quam' in Latin) to express 'as possible,' *e.g.* *ὡς πλεῖστοι* 'as many as possible'; *ὅτι* or *ὡς τάχιστα* 'as quickly as possible.'

οἰκεῖος -α -ον belonging to the home.
ἐπιθυμέω I desire (*gen.*).
χαίρω I rejoice, *fut.* χαιρήσω, *aor.* ἐχάρην.
ἡ ἡδονή pleasure.
ἀνα-τείνω stretch up *or* out.
χαῖρε hail *or* farewell.

Ex. 60. 1. ἀεὶ δίωκε τὴν ἀρετήν, ὦ παῖ, καὶ φεῦγε τὰς κακὰς ἡδονάς. τοὺς μὲν θεοὺς φοβοῦ, τοῖς δὲ νόμοις πείθου.

2. μετὰ δὲ ταῦτα ὁ Ξενοφῶν ἔφη, Καὶ ὅτῳ δοκεῖ ταῦτα, ἀνατεινάτω τὴν χεῖρα· καὶ ἀνέτειναν ἅπαντες. Καὶ ὅστις ὑμῶν τοὺς οἰκείους ἐπιθυμεῖ ἰδεῖν μεμνήσθω ἀνὴρ ἀγαθὸς εἶναι· ὅστις δὲ ζῆν βούλεται πειράσθω νικᾶν.

3. ὁ Λεωνίδας, τοῦ Ξέρξου γράψαντος Πέμψον τὰ ὅπλα, ἀπεκρίνατο, Ἐλθὼν λαβέ.

4. ἴσθι ἀγαθή, ὦ φίλη κόρη (girl), καὶ ἔστω σοφὴ ἥτις δύναται.

5. παρασκευασθέντων αἱ τριήρεις εὐθύς.

6. ὦ παῖ, χρῶ ᾧ ἔχεις νῷ καὶ πειρῶ μαθεῖν ταῦτα.

7. Χαῖρε, ὦ φίλε, ἔφη, καὶ γράφε πρὸς ἐμὲ πολλάκις.

8. τειχιζόντων τὴν πόλιν οἱ πολῖται.

9. λέγω γὰρ τούτῳ, Πορεύθητι, καὶ πορεύεται· καὶ ἄλλῳ, Ἔρχου, καὶ ἔρχεται καὶ τῷ δούλῳ μου Ποίησον τοῦτο, καὶ ποιεῖ.

10. σωφρονέστερον ἔσται ἀπελθεῖν ὡς τάχιστα.

Ex. 61. 1. Honour thy father and thy mother and always obey them.

2. Let the doors be shut and let all the household be present.

3. O men of Athens, said he, remember the deeds of (things done by) your fathers in the war against the Persians and try to be yourselves worthy of them. So let everyone think that he is fighting not only for (ὑπέρ) himself but for his country (fatherland).

4. Ransom this man and let him be loosed from prison as quickly as possible.

5. Let them therefore all die for they have done things unworthy of our city.

6. Let the house be sold and all that (ὅσα) is in it.

7. Let these men be honoured since they have saved the city.

8. Let the young men and maidens (παρθένος) rejoice and let them dance in the market-place.

9. Let all those who wish to cease from work put up their hands.

10. But tell me, "Who are these men and what are they trying to do?"

11. Knock at the door often for the poet is said not to hear well (ὀξέως *adv. of* ὀξύς 'sharp').

12. You are more fortunate than your brother.

THIRD DECLENSION STEMS ENDING IN -ευ, -ου, -αυ

These are irregular as the υ drops before a vowel and one of the vowels may be lengthened. The nouns in -ευς express an agent.

	king	ox, cow	ship	old woman
Sing. Nom.	ὁ βασιλεύς	ὁ ἡ βοῦς	ἡ ναῦς	ἡ γραῦς
Voc.	βασιλεῦ	βοῦ	ναῦ	γραῦ
Acc.	βασιλέα	βοῦν	ναῦν	γραῦν
Gen.	βασιλέως	βοός	νεώς	γραός
Dat.	βασιλεῖ	βοΐ	νηΐ	γραΐ
Plur. N. V.	βασιλῆς	βόες	νῆες	γρᾶες
Acc.	βασιλέας	βοῦς	ναῦς	γραῦς
Gen.	βασιλέων	βοῶν	νεῶν	γραῶν
Dat.	βασιλεῦσι(ν)	βουσί(ν)	ναυσί(ν)	γραυσί(ν)

THE SUBJUNCTIVE MOOD

The connecting vowel of the Indicative is lengthened and the ι written subscript.

Active				*Mid. and Passive*			
Sing.	λύω	*Plur.*	λύωμεν	*Sing.*	λύωμαι	*Plur.*	λυώμεθα
	λύῃς		λύητε		λυῇ		λύησθε
	λύῃ		λύωσι(ν)		λύηται		λύωνται

These endings apply to all tenses Active, Middle and Passive but the Perf. Passive is formed as in Latin—Perf. Partic. + Subj. of εἰμί ('sum').

λελυμένος ὦ, ᾖς, ᾖ, λελυμένοι ὦμεν, ἦτε, ὦσι(ν).

The Aor. Pass. Subj. of all verbs has the Act. endings, λυθῶ -ῇς -ῇ etc.

The Subj. of verbs in -όω: δηλῶ, δηλοῖς, δηλοῖ, δηλ-ῶμεν, -ῶτε, -ῶσι.

PROHIBITIONS (NEGATIVE COMMANDS)

μή with *Present* Imperative: μὴ ποίει τοῦτο 'Don't do this' (habitually): 'Don't go on doing this.'

μή with *Aorist* Subjunctive: μὴ ποιήσῃς τοῦτο 'Don't do this.' See p. 58 (*c*).

Exhortations. The Subj. is used for the 1st person (but Imperative always for the 3rd); ἄγε or φέρε are sometimes prefixed.

(ἄγε) χορεύωμεν '(Come) let us dance' (but χορευόντων 'Let them dance').

DELIBERATIVE QUESTION

A question expressing doubt or bewilderment is put in the Subj. τί λέγωμεν; 'What are we to say?' πότερον πορευθῶ ἢ μή; 'Am I to set out or not?'

ἔτι still, yet.
ὁ ἱππεύς horseman.
τήμερον (σήμερον) to-day.
βοηθέω go to help (*dat.*).
σιγάω I am silent.
ὁ ἑρμηνεύς interpreter.
ἡ πύλη gate.
μηδέποτε / μὴ...ποτέ never.
μήτε...μήτε neither...nor.

Ex. 62. 1. μὴ φύγωμεν, ὦ ἄνδρες, ἀλλ' ἀποθάνωμεν μαχόμενοι ὑπὲρ τῆς πατρίδος. 2. μὴ κλείσητε τήμερον τὰς τῆς πόλεως πύλας οἱ γὰρ ἱππῆς ἔτι διώκουσι τοὺς πολεμίους. 3. μηδέποτε λεῖπε τὴν τάξιν ἐν μάχῃ. 4. ὦ βασιλεῦ, βοηθῶμεν τοῖς ἱππεῦσιν· μὴ νικηθέντων ὑπὸ τῶν πολεμίων. 5. ἄγε ἐλαύνωμεν τὰς βοῦς εἰς τὴν ἀγοράν. 6. εἴπωμεν ἢ σιγῶμεν ἢ τί δράσομεν;

7. *The Lord's Prayer.*

ὧδε οὖν προσεύχεσθε ὑμεῖς·

Πάτερ ἡμῶν ὁ ἐν τοῖς οὐρανοῖς· ἁγιασθήτω τὸ ὄνομά σου· ἐλθέτω ἡ βασιλεία σου· γενηθήτω τὸ θέλημά σου, ὡς ἐν οὐρανῷ καὶ ἐπὶ γῆς· τὸν ἄρτον ἡμῶν τὸν ἐπιούσιον δὸς ἡμῖν σήμερον· καὶ ἄφες ἡμῖν τὰ ὀφειλήματα ἡμῶν, ὡς καὶ ἡμεῖς ἀφήκαμεν τοῖς ὀφειλέταις ἡμῶν· καὶ μὴ εἰσενέγκῃς ἡμᾶς εἰς πειρασμόν, ἀλλὰ ῥῦσαι ἡμᾶς ἀπὸ τοῦ πονηροῦ.

8. ἔλαβον δέκα τῶν νεῶν αὐτοῖς ἀνδράσιν. 9. μὴ φοβηθῇς τὸν κύνα. 10. πέμψας ἑρμηνέα παρὰ τοὺς στρατηγοὺς τῶν Ἑλλήνων ἐκέλευσε τὴν φάλαγγα ἀναχωρῆσαι. 11. μέμνησό μου καὶ πέμπε ἐπιστολὴν κατὰ τὸν μῆνα ἕκαστον. 12. ἄγε δή, ὦ φίλε, νῦν μοι βοήθει.

Ex. 63. 1. Let us remain. Are we to remain? Shall we remain? Let us not remain. 2. Do not go out of the city. Remain in your own house. 3. Let the horsemen go to help the king. 4. Let us all pray to the gods on behalf of ourselves and our friends. 5. Let him speak to us through the interpreter. 6. What, then, are we to do? For the enemy are already at the gates. 7. Let all these (events) be announced to the people and let the citizens vote concerning them. 8. Do not prepare that ship for a truce will be (made) as quickly as possible. 9. Let us send a herald to the ship with *an* interpreter. 10. Be more prudent, Your Majesty, and retreat while it is still possible. 11. Do not put these men to death for it is not expedient although they have injured us but let them be sold and become slaves. 12. Let us send as many ships as possible into the harbour. 13. Try to be more worthy of your father.

IRREGULAR COMPARISON OF ADJECTIVES
-ίων, -ιστος. ADVERBS -ιον, -ιστα

ἐχθρός	hostile	ἐχθ-ίων	ἔχθ-ιστος
αἰσχρός	shameful	αἰσχίων	αἴσχιστος
ταχύς	swift	θάττων	τάχιστος
ἡδύς	pleasant	ἡδίων	ἥδιστος

The following have irregularities of stem:

ἀγαθός	good	ἀμείνων	ἄριστος
(εὖ *adv.* well)		βελτίων	βέλτιστος
		κρείττων	κράτιστος
κακός	bad *or*	κακίων	κάκιστος
	cowardly	χείρων	χείριστος
ὀλίγος	little	ἐλάττων	ἐλάχιστος
	pl. few	ἥττων	(ἥκιστα *adv.*)
πολύς	much, *pl.* many	πλείων (πλέων)	πλεῖστος
ῥᾴδιος	easy	ῥᾴων	ῥᾷστος
μικρός, σμικρός	small	(σ)μικρότερος	(σ)μικρότατος
		μείων	
ἀλγεινός	painful	ἀλγίων	ἄλγιστος

Adjectives expressing size usually form adverbs by taking the neuter instead of -ως: πολύ, μικρόν, ὀλίγον, σφόδρα ('exceedingly').

Note: μάλα ('much'), μᾶλλον ('more'), μάλιστα ('most' *or* 'especially') is used as adv. of μέγας.

Adverbs in -ω (many formed from prepositions) are compared thus: ἄνω 'up,' ἀνωτέρω, ἀνωτάτω.

So κάτω 'down,' εἴσω 'inside,' ἔξω 'outside,' πόρρω 'far.'

PURPOSE (FINAL CLAUSE)

(1) ἵνα, ὡς, *or* ὅπως with Subjunctive. Negative μή. (Optative can be used if the main verb is past.)

(2) Future participle (preceded by ὡς if the verb does not express motion or if the purpose is only a presumed one).

(3) Future indic. with ὅστις after verbs of sending, choosing, and finding.

(1) τοῦτο ποιεῖ ἵνα πλείω χρήματα λάβῃ
He does this to get more money ('ut capiat').

(2) ἦλθεν ὀψόμενος τοὺς ἀγῶνας ('visurus')
He came to see the sports.

(3) πέμψον ἄνδρα ὅστις ἀγγελεῖ ταῦτα
Send a man to announce this (who shall announce).

ἐπαινέω I praise.
διάγω pass the time, live.
τὸ δένδρον tree, *dat. pl.* δένδρεσι.
ὁ οἶνος wine.
τὸ μέλι μέλιτ- honey.
ἴσος *adj.* equal.

Ex. 64. 1. κατέκαυσαν ὡς πλείστας τῶν νεῶν ἵνα μὴ οἱ Πέρσαι εἰς τὰς νήσους εἰσπλέωσιν.

2. πόσῳ κάλλιόν ἐστιν ἀποθανεῖν μαχόμενος ἄριστα ὑπὲρ τῆς πατρίδος ἢ διάγειν αἴσχιστα τὸν βίον ἐν τῇ κακίστῃ δουλείᾳ.

3. ἔγραψεν ὁ Πίνδαρος, ἀρχόμενος λυρικῆς τινὸς ᾠδῆς, Ἄριστον μὲν ὕδωρ· οἱ δὲ πλεῖστοι τῶν ποιητῶν, μάλιστα ὁ Ἀνακρέων, ἐπαινεῖν φιλοῦσι τὸν οἶνον, καίπερ ἴσον ἴσῳ κε-κρα-μένον (mixed), ὡς καὶ ὁ Ὅμηρος λέγει ἐν τῇ Ἰλιάδι περὶ τῶν ἡρώων.

4. ὅσῳ πλείους οἱ πολέμιοι εἰσίν, τοσούτῳ μείζων ἡ δόξα (glory) ὑμῖν ἔσται νικήσασιν αὐτούς. ἆρ' οὐχ οὕτως καὶ ὑμῖν δοκεῖ;

5. ὁ δὲ Ἐτεόνικος ἔμεινε παρὰ ταῖς πυλαῖς ὡς συγκλείσων τὰς πύλας καὶ τὸν μοχλὸν (bar) ἐμ-βαλῶν.

6. ὑπέσχετο γὰρ κομιεῖν (*p.* 54, *N.B.*) πλείους καὶ μείζους ναῦς ὥστε ἐλπίζομεν δυνήσεσθαι νικᾶν τοὺς Πέρσας ῥᾷστα κατὰ θάλατταν.

7. θάψων γὰρ ἥκω Καίσαρ' οὐκ ἐπαινέσων.

Ex. 65. 1. Let us march at once that we may more easily reach (ἀφικνέομαι εἰς) the city before night; for it is far distant.

2. We at least consider (it) of more importance to win a few victories than to write many fine poems.

3. Send me more and better books (βιβλίον) that I may learn as much as possible about music and mathematics (Ex. 12).

4. What is sweeter than honey? What is stronger than a lion?

5. The trees that you have in this land are smaller and fewer than those in ours. Tell me why this is so (ἔχω, *p.* 22).

6. The sooner (ὅσῳ θᾶττον) you reply (*fut.*) to my question ('to me asking') the better it will be for you.

7. Of all the cities which (*p.* 20 (2)) I saw while I was travelling I think Neapolis is the most beautiful, especially at night.

8. They went away as if to sail (ὡς *with fut. part.*) home.

9. There is nothing that I will not do to help you.

(1) An event in past time is expressed by the Indicative with

ὅτε = when, at the time when (purely temporal).
ἐξ οὗ, ἀφ' οὗ, ὡς = since (from the time when).
ἐπεὶ τάχιστα = as soon as.
ὡς, ἐπεί, ἐπειδή, ἡνίκα = when.
ἕως, μέχρι (οὗ), ἄχρι, ἔστε = until.
ἕως, ἐν ᾧ, ἐν ὅσῳ, ὅσον χρόνον = while.
ἐπειδή, ὡς = after.
πρίν = before (see p. 89).
In the time of, in the reign of = *ἐπί* with genitive.
ἐπὶ τοῦ Σόλωνος = in the days of Solon.
In many sentences only the participle is used.

(2) Indefinite time is expressed in Primary Sentences by *ὅταν, ἐπειδάν, ἐπήν*, 'whenever'; *ἐάν*, 'if ever,' and the Subjunctive.

For indefinite time in Historic Sentences, see p. 66 (3).

(1) *ἐπειδὴ προσ-έμιξαν, οἱ ἐπιβάται ἐπειρῶντο ἐπιβαίνειν*
When (the ships) came close, the marines tried to board.

Here a definite event, which happened once, is meant.

(2) *ἐπειδὰν προσμίξωσιν, οἱ ἐπιβάται πειράσονται ἐπιβαίνειν*
When(ever) they come close, the marines will try to board.

'When' is often used loosely for 'whenever' in English.

(1) *καὶ ταῦτα ἐποίουν μέχρι σκότος ἐγένετο*
And this they went on doing till darkness came.
ὡς αὐτοὺς οὐκ ἔπεισεν, ἀπ-έπλευσεν
When he failed to convince them, he sailed away.

(2) *ἐπειδὰν τάχιστα ἱππεύειν μάθῃς, διώξει τὰ θηρία*
As soon as ever you have learnt to ride, you shall hunt the beasts.
μέχρι δ' ἂν ἐγὼ ἥκω (subj.), *αἱ σπονδαὶ μενόντων*
Until I come, let the truce remain in force.

N.B. The same rule applies to relatives; when the clause is indefinite they will be used with *ἄν* and the subjunctive, the *ἄν* corresponding to the English word 'ever'—'whoever,' 'whatever': *πράττουσιν ἃ ἂν βούλωνται* = they always do whatever they like.

In Temporal Sentences indefinite time means future time; or a repeated occurrence without reference to time.

ὁ ἥλιος sun. χαρίζομαι show favour, please (*dat.*).
συμβουλεύομαι consult together. πρότερος former, first of two.

Ex. 66. 1. ἦν ποτὲ χρόνος ὅτε θεοὶ μὲν ἦσαν, θνητὰ δὲ γένη οὐκ ἦν.

2. ἄγγελλε τῷ Μαρδονίῳ ὅτι οἱ Ἀθηναῖοι λέγουσιν, Ἕως ἂν ὁ ἥλιος τὴν αὐτὴν ὁδὸν ἴῃ (goes) ᾗ καὶ νῦν, οὐδέποτε ἡμᾶς ποιήσειν ἃ Ξέρξης κελεύει.

3. ἃ ἂν κελεύῃ ὁ κριτής, ταῦτα ποίει, καὶ ἀσφαλὴς ἔσει.

4. ὅταν δέῃ (*subj. of* δεῖ) πείθειν τινὰ τῶν φίλων, χρήματα πέμπει, ἵνα ἐθέλῃ ἐκεῖνος ποιεῖν ἃ αὐτὸς βούλεται.

5. πολλὰ δὴ ἔτη ἐστίν, ἐξ οὗ ταῦτα ἐγένετο.

6. ἐπειδὰν οἱ σύμμαχοι πάρωσιν, δυνησόμεθα μάχεσθαι τοῖς πολεμίοις ἐπ' ἴσῃ καὶ ὁμοίᾳ.

7. ταῦτ' οὖν ἔγωγε νομίζω εἶναι ἃ δεῖ ποιεῖν, ὅτῳ δ' ἂν δοκῇ ταῦτα οὐκ ἀσφαλῆ, παρελθὼν λεγέτω ὡς συμβουλευσώμεθα περὶ τοῦ πράγματος.

8. ἐὰν ἴδῃ τινὰ κακῶς ἔχοντα, τοῦτον ἀεὶ ὠφελεῖν πειρᾶται.

9. τοῦ δὲ στρατηγοῦ ἀποθανόντος οἱ στρατιῶται ἀθύμως δὴ εἶχον ὡς ἐν πολεμίᾳ χώρᾳ ὄντες.

10. ὁ ἐν Δελφοῖς θεός, ὅταν τις ἐρωτᾷ πῶς δεῖ τοῖς θεοῖς χαρίζεσθαι, ἀποκρίνεται, Νόμῳ πόλεως.

11. ἐπειδὴ οὖν καὶ πέμψαντες πρέσβεις οὐκ ἔπειθον παύσασθαι τοῦ πολέμου τοὺς ἐν ταῖς νήσοις, ἐδόκει τοῖς Ἀθηναίοις ἀποστέλλειν τὰς τριήρεις τάς τε ἑαυτῶν καὶ τὰς τῶν συμμάχων.

Ex. 67. 1. In the reign of Xerxes the Greeks conquered the Persians both by land and sea. 2. As soon as he saw us coming he fled from the city into the mountains. 3. It is a long time since all this happened. 4. You must do whatever the king orders. 5. Let us send someone to inform the general what is happening. 6. The men who fought so nobly in this battle ought to be honoured by all the citizens. 7. The-man-who (ὅστις) tries to please everybody, often pleases no one. 8. When there is an earthquake everybody rushes into the market-place that he may not be injured. 9. When Cleon was general many thought that we should be defeated. 10. We shall cease from fighting until the allies arrive. 11. What happened after the battle has been explained (δηλόω) in the former book (βιβλίον). 12. When they arrived in the city they found many of the allies dying of disease; and so (ὥστε) it seemed good to them not to go against the enemy. 13. There was once a time when men did not exist. 14. Do not tell anybody what we have done. 15. When everything has been prepared send a messenger, that I may accompany you.

THE OPTATIVE MOOD

λύω *Active*

εἰμί ('sum')	*Pres., Fut. and Perf.*	*Aor.*	*Mid. and Pass. Pres.*
Sing. εἴην	-οιμι	-αιμι	-οίμην
εἴης	-οις	-ειας	-οιο
εἴη	-οι	-ειε(ν)	-οιτο
Plur. εἶμεν	-οιμεν	-αιμεν	-οίμεθα
εἶτε	-οιτε	-αιτε	-οισθε
εἶεν	-οιεν	-ειαν	-οιντο

but ἐλυσάμην and verbs in -αμαι will form -αιμην, -αιο, -αιτο etc.

The Aor. Pass. follows εἰμί, *e.g.* λυθείην. The Perf. Act. is usually λελυκὼς εἴην and the Perf. Pass. λελυμένος εἴην.

Contracted Verbs follow εἰμί: τιμαοίην—τιμῴην, τιμῴης etc.; so φιλοίην, δηλοίην, but in other tenses they follow λύω.

IMPORTANT USES OF THE OPTATIVE

(1) To express a wish for the future; often introduced by εἴθε, εἰ γάρ, or εἰ: εἴθε σωζοίμεθα 'O that we might be saved!'

(εἰ γὰρ) μὴ γένοιτο 'May it not happen!' Neg. μή.

(2) With ἄν to express a doubtful future. Thus δυνήσομαι 'I shall be able,' δυναίμην ἄν '(perhaps) I might be able.'

οὐκ ἂν νικῴης 'you probably won't win.' Neg. οὐ.

(3) With ὅποτε, ἐπειδή, εἰ, in Historic Time to express a frequent or indefinite event: '*whenever* he saw me, he used to avoid me' ὅποτε ἴδοι με, ἔφευγεν.

Also with relatives with indefinite antecedent.

οὓς ἴδοι εὐτάκτως πορευομένους ἐπῄνει

He used to praise those whom he saw marching well.

αἱρέω and ἁλίσκομαι.

αἱρέω 'I take,' αἱρήσω, εἷλον, ᾕρηκα.

αἱροῦμαι 'I choose,' αἱρήσομαι, εἱλόμην, ᾕρημαι.

ἁλίσκομαι 'I am taken,' ἁλώσομαι, ἑάλων, ἑάλωκα.

αἱροῦμαι 'I am chosen,' αἱρεθήσομαι, ᾑρέθην, ᾕρημαι,

but in compounds αἱρέω has its usual meaning in the Passive.

εἷλον, ἕλε, ἕλω, ἕλοιμι, ἑλεῖν, ἑλών 'I took.'

εἱλόμην, ἑλοῦ, ἕλωμαι, ἑλοίμην, ἑλέσθαι, ἑλόμενος 'I chose.'

ἑάλων, ἅλωθι, ἁλῶ, ἁλοίην, ἁλῶναι, ἁλούς 'I was taken.'

ἑάλων, ἑάλως, ἑάλω ἑάλωμεν, ἑάλωτε, ἑάλωσαν, or ἥλων κ.τ.λ.

γυμνάζω train, exercise.
θηρεύω, θηράω hunt, catch.
αὖ, αὖθις again.
οὐδέ not even.
ὁ θήρ, τὸ θηρίον wild beast.
δια-τελέω continue (*with partic.*).

Ex. 68. 1. τούτων δὲ τῶν κακῶν πότερον ἑλώμεθα; πότερον δὴ ἕλοιο ἂν αὐτός;

2. Περικλῆς ὁ Ξανθίππου ᾑρέθη στρατηγὸς δέκατος αὐτός.

3. οἱ Ἀθηναῖοι πέντε ναῦς εἷλον καὶ μίαν τούτων αὐτοῖς ἀνδράσιν.

4. ὦ παῖ, γένοιο πατρὸς εὐτυχέστερος, τὰ δ' ἄλλα ὅμοιος, καὶ γένοιο ἂν οὐ κακός.

5. ὁ γὰρ Κῦρος ἐθήρευε ἐκεῖ τὰ θηρία ἀφ' ἵππου ὅποτε γυμνάσαι βούλοιτο ἑαυτόν τε καὶ τοὺς ἵππους.

6. βασιλεὺς αἱρεῖται, οὐχ ἵνα ἑαυτοῦ καλῶς ἐπι-μελῆται ἀλλ' ἵνα καὶ οἱ ἑλόμενοι δι' αὐτὸν εὖ πράττωσιν.

7. τοῦτό γε κἂν (καὶ ἂν) παῖς ἐν τῷ γυμνασίῳ γνοίη (know).

8. μή μοι ἃ βούλομαι γένοιτο ἀλλ' ἃ συμφέρει.

9. ὅταν πόλις ἁλῷ, τῶν ἑλόντων ἔστι (belong) καὶ τὰ σώματα τῶν ἐν τῇ πόλει καὶ τὰ χρήματα.

10. ἴσως δ' ἂν εἴποι τις· πῶς ἂν τοῦτο ποιεῖν δυναίμεθα ἅτε οὐκ ἔχοντες οὐδὲ μίαν ναῦν;

11. οἱ ἐν ταύτῃ τῇ πόλει ἁλόντες ἦσαν πόλλῳ πλείους τῶν ἑλόντων.

12. τῆς δὲ φάλαγγος λυθείσης οἱ μὲν ἀπέθανον οἱ δὲ πλεῖστοι ἑάλωσαν.

13. οὐδεὶς τοιοῦτο ἂν ἐπαινοίη.

14. τίς οὐκ ἂν ἕλοιτο τὸ πεπαιδεῦσθαι ἀντὶ τοῦ ἀπαίδευτος διατελεῖν πάντα τὸν βίον;

Ex. 69. 1. Perhaps you would like to hear what happened that night in the city. 2. So whenever he comes to Athens we receive him into our house as-a-guest, since we ourselves were always entertained (ξενίζω) by him whenever we were in *his* country. 3. When the city was taken, they took away (ἀφαιρέω) all the money that they could find. 4. Would that my brother might be chosen general! 5. We shall therefore put you to death that no one (ἵνα μή τις) else may do the same (things). 6. I at least will do whatever the king commands. 7. Which-of-the-two slaves did you choose? 8. How then could we trust again a-man-who-has-injured us often in former years? 9. He said he wished to hunt lions but whenever he saw a lion coming he-used-to-run-away. 10. I have been chosen general with three others. 11. The tyrant used to manage (πράττω) that no one might become more powerful than himself. 12. Even a slave would be braver than this man.

PART III

STRONG AORISTS ACTIVE IN -ων, -ην, -ᾱν, -ῡν

	Aor. Indic.	*Imperative*	*Subj.*	*Opt.*	*Infin.*
Sing.	ἔγνων		γνῶ	γνοίην	γνῶναι
	ἔγνως	γνῶθι	γνῷς	γνοίης	*Partic.*
	ἔγνω	γνώτω	γνῷ	γνοίη	γνούς
Plur.	ἔγνωμεν		γνῶμεν	γνοῖμεν	*Stem*
	ἔγνωτε	γνῶτε	γνῶτε	γνοῖτε	γνόντ-
	ἔγνωσαν	γνόντων	γνῶσι(ν)	γνοῖεν	

Similarly ἔβην 'I went' substituting η for ω and α for ο. Part. βάς (βαντ-). Subj. βῶ, βῇς, βῇ, κ.τ.λ.

Similarly ἀπ-έδρᾱν 'I ran away' substituting ᾱ for ω and α for ο. Part. ἀποδράς.

Similarly ἔδυν 'I sank' which has υ throughout. Part. δύς.

ἔγνων is the aor. act. of γιγνώσκω 'I get to know,' γνώσομαι, ἔγνων, ἔγνωκα, ἔγνωσμαι, ἐγνώσθην.

οἶδα, εἴσομαι, ᾔδη.

οἶδα perfect in form, present in meaning, Stem εἰδ- 'I know.'

Pres.				*Imperf.*			
Sing.	οἶδα	*Plur.*	ἴσμεν	*Sing.*	ᾔδη	*Plur.*	ᾖσμεν
	οἶσθα		ἴστε		ᾔδησθα		ᾖστε
	οἶδε(ν)		ἴσασι(ν)		ᾔδει(ν)		ᾔδεσαν (ᾖσαν)

Imperative ἴσθι, ἴστω, ἴστε. *Subj.* εἰδῶ. *Opt.* εἰδείην.
Participle εἰδώς, εἰδυῖα, εἰδός. *Stem* εἰδότ-. *Infin.* εἰδέναι.

Verbs of knowing and perceiving are usually followed by a participle (instead of the infin. or the ὅτι and indic. construction); many similar verbs take either construction; ἀκούω, ἀγγέλλω, μέμνημαι, πυνθάνομαι, 'ascertain' etc. When the participle refers to the subject it will be in the nom. case; otherwise it will agree with the object in whatever case the verb requires, *e.g.* οἶδα ἀδικήσας 'I know I did wrong.'

οἶδα αὐτὸν ἀδικήσαντα 'I know that he did wrong.'

ἆρ' οὐ μέμνησαι ταῦτα εἰπών; 'don't you remember saying that?'

ἡδύ ἐστι πυνθάνεσθαι τοὺς φίλους εὖ πράττοντας 'it is pleasant to know that one's friends are faring well.'

ᾔσθοντο οἱ στρατηγοὶ οὐ κατορθοῦντες, τοὺς δὲ στρατιώτας ἀχθομένους 'the generals realised that they (themselves) were not succeeding, while the soldiers were becoming annoyed.'

ἡ σωτηρία safety.
ὁ ἀγών contest.
κατα-γιγνώσκω condemn (*gen. of person*).
συγ-γιγνώσκω pardon (*dat.*).
ἡ συμφορά disaster.
ὁμολογέω I confess.
ἀνα-γιγνώσκω read.
μετα-γιγνώσκω repent.

Ex. 70. 1. ταῦτα δὲ λέγω ἵνα εἴδητε δικαίως κατεγνωκότες αὐτῶν.
2. ἐν δὲ τούτῳ οἱ Ἕλληνες ἔμειναν ἐν τῷ στρατοπέδῳ· οὐ γὰρ ᾔδεσαν τὸν Κῦρον τεθνηκότα.
3. ὁ δὲ λύσας τὴν ἐπιστολὴν ἀνέγνω.
4. ὦ ἄνθρωπε, γνῶθι σεαυτόν.
5. ταχέως οὖν εἰσόμεθα πόση ἐστὶν ἡ συμφορά, ὁρῶ γὰρ ἄγγελον ἤδη προσερχόμενον.
6. οἱ κύνες βαΰζουσιν ὃν ἂν μὴ γιγνώσκωσιν.
7. μὴ θαυμάσῃς εἰ πάντες ἴσασι ταῦτα· οὐ γὰρ δυνατὸν ἦν κρύπτειν πολλὰς ἡμέρας.
8. ὅταν μέν τις τὸν ἀδικήσαντα ἀδικήσῃ, τί μεῖζον ἢ κατὰ τὸν ἐχθρὸν αὐτὸν ἐποίησεν; συγγνοὺς δὲ τῷ ἀδικήσαντι τὴν ἀδικίαν, βασιλικόν τι ἐποίησεν· βασιλέως γάρ ἐστι τὸ συγγνῶναι.
9. εὖ γὰρ ἴστε, ὦ ἄνδρες, μαχόμενοι οὐ μόνον περὶ τῆς νίκης ἀλλὰ καὶ περὶ τῆς σωτηρίας τῆς τε πόλεως καὶ ὑμῶν αὐτῶν.
10. ἐπεὶ δ' οὖν ὑμεῖς ἢ ἴστε οὐδὲν ἢ οὐ τολμᾶτε λέγειν, ἀλλ' ἔγωγε λέξω ὡς εὖ ἐπιστάμενος.
11. ὡμολόγουν τοὺς εὖ ἐμὲ δράσαντας κακῶς δράσας.
12. διῆλθον οὖν ἅπασαν τὴν χώραν ἵν' ἔχοιεν τῷ βασιλεῖ εἰπεῖν ὅτι οὐδεὶς σφᾶς ἀποπεφευγὼς εἴη.
13. ἆρ' οὐκ ᾔσθου αὐτὸν εἰδότα ταῦτα;

Ex. 71. 1. He soon perceived that he was no longer honoured by the men in the islands but that many suspected him. 2. Changing his mind (*aor. part.*) he pardoned those whom he had condemned to death (*acc.*). 3. Don't you know that you are injuring the state by doing this? 4. Know well, o judge, that you have condemned an innocent (ἀναίτιος) man. 5. When he perceived that the house was on fire, instead of trying to do something useful, he sat down and wept. 6. Who does not know that virtue is better than much gold? 7. I have sent a messenger that they may know how great the disaster is. 8. Be brave, my son, remembering that *you* are an Athenian and that *these* men are barbarians. 9. When they heard that the king was dead so far from (οὐχ ὅπως) attacking (*aor. ind.*) they fled across the plain throwing away their arms. 10. May you never repent of (περὶ) this deed! 11. Whenever he perceived that a struggle was taking place, he desired to have a share (μετέχω *with gen.*) in it. 12. They confessed that they had taken the money which we hid in the house.

GOING AND COMING

These ideas are rendered by a variety of verbs which are defective and supply each other's deficiencies (cf. 'aller' in French, 'go' in English).

Pres.	*Ind.*	*Imp.*	*Subj.*	*Opt.*	*Infin.*	*Partic.*
	ἔρχομαι	ἴθι, ἴτω	ἴω	ἴοιμι	ἔρχεσθαι	ἐρχόμενος
	or εἶμι	ἴτε, ἰόντων			*or* ἰέναι	*or* ἰών
Aor.	ἦλθον	ἐλθέ	ἔλθω	ἔλθοιμι	ἐλθεῖν	ἐλθών

Imperf. ᾖα

Fut. εἶμι or ἥξω.

Perf. ἐλήλυθα, βέβηκα, ἥκω 'I am come,' οἴχομαι 'I am gone.'

Note. The two last are present in form, perfect in meaning.

Pres.				*Imperf.*			
Sing.	εἶμι	*Plur.*	ἴμεν	*Sing.*	ᾖα *or* ᾔειν	*Plur.*	ᾖμεν
	εἶ		ἴτε		ᾔεισθα		ᾖτε
	εἶσι(ν)		ἴασι(ν)		ᾔει		ᾔεσαν (ᾖσαν)

Compounds: by prefixing prepositions various meanings can be expressed:

ἀπῆλθεν	κατιέναι	ἐξελθεῖν	ἄπιθι	προσίωμεν
he departed	to descend	to go out	begone!	let us approach

βαίνω, βήσομαι, ἔβην, βέβηκα is used in prose chiefly in compounds:

ἀναβαίνω	to go up, go inland.
διαβαίνω	cross (river or plain).
ἐμβαίνω, εἰσβαίνω	go on board ship.
ὑπερβαίνω	cross (mountain or wall).
ἐκβαίνω, ἀποβαίνω	disembark (intr.).
παραβαίνω	transgress.

The transitive form is βιβάζω, *e.g.* ἐκ-βιβάζω 'to disembark' (troops). Attic Fut. βιβῶ -ᾷς -ᾷ.

Note also the imperative δεῦρο, pl. δεῦτε 'come hither.'
'arrive' ἀφικνέομαι, ἀφίξομαι, ἀφικόμην, ἀφῖγμαι.
'advance' ἐπιχωρέω. 'retreat' ἀναχωρέω.
'walk' βαδίζω (fut. βαδιῶ). ἡ ὁδὸς φέρει...'the road goes....'

ἀπο-διδράσκω, ἀπο-δράσομαι, ἀπ-έδρᾱν, ἀπο-δέδρακα run away.

ἑκών -οῦσα -όν willing(ly). ἄκων unwillingly.

οἴκοι at home. οἴκαδε home(wards).

ὑψηλός -ή -όν high. συγ-καλέω call together.

νὴ τοὺς θεούς by the gods! οὐ μὰ τοὺς θεούς No, by heaven!

ἀπο-λείπω leave, abandon, *aor.* ἔλιπον, *perf.* λέλοιπα.

Ex. 72. 1. ἐπεὶ δὲ οἱ πρέσβεις ἧκον, παρελθὼν ὁ Κλέων εἶπε τάδε. 2. ᾤχετο ἀπιὼν οἴκαδε. 3. τί βουλόμενοι δεῦρο ἥκετε; ἀλλ' οὐ χαίροντες ἄπιτε. 4. ἡμεῖς γὰρ πολλοὺς καὶ μεγάλους ποταμοὺς διαβάντες, καὶ ὑπερβάντες δεινὰ καὶ ὑψηλὰ ὄρη, τέλος ἀφίγμεθα πρὸς τὸν Βαβυλῶνα. 5. καθ' ἡμέραν μὲν οἴκοι παρέμεινε νυκτὸς δὲ ἐξῄει εἰς τὴν πόλιν. 6. ἐπειδὴ οὖν ἀφίκοντο ἐμβιβασάμενος ἄλλους εἰς τὰς ναῦς ἔπλει ὡς τάχιστα εἰς τὸν λιμένα οὐ πολὺ ἀπέχοντα. 7. ἐνταῦθ' ἔμειναν ἡμέρας ἑπτά· καὶ Ξενίας ὁ Ἀρκάς, στρατηγός, καὶ Πασίων ἐμβάντες εἰς πλοῖον ἀπέπλευσαν. Κῦρος δὲ συγκαλέσας τοὺς στρατηγοὺς εἶπεν· Ἀπολελοίπασιν ἡμᾶς Ξενίας καὶ Πασίων· ἀλλ' εὖ γε μέντοι ἐπιστάσθων ὅτι οὔτε ἀποδεδράκασιν· οἶδα γὰρ ὅπῃ οἴχονται· οὔτε ἀποπεφεύγασιν· ἔχω γὰρ τριήρεις ὥστε ἑλεῖν τὸ ἐκείνων πλοῖον. ἀλλά, μὰ τοὺς θεούς, οὐκ ἔγωγε αὐτοὺς διώξομαι. ἀλλ' ἰόντων, εἰδότες ὅτι κακίους εἰσὶ περὶ ἡμᾶς ἢ ἡμεῖς περὶ ἐκείνους. 8. ὅποτε βαδίζειν πειρῷτο, κατέπιπτεν. 9. Κυμαῖός τις περὶ κλίμακός (ladder) τινος πυθόμενος ὅτι τοῖς ἀναβαίνουσιν ἔχει δώδεκα βάθρα (rung), ἠρώτησεν εἰ καὶ τοῖς καταβαίνουσι τοσαῦτά ἐστιν.

Ex. 73. 1. The soldiers however at first refused to go; but when Cyrus promised (*gen. abs.*) greater pay (μισθός *masc.*) they changed their minds and went willingly. 2. As soon as they saw troops coming down from the mountains they ran away homewards. 3. Let no one (μηδείς) go in or (μήτε) come out while I am absent (ἄπειμι). 4. When all had come together, it seemed good to the general to order some to go in search of (ἐπί *with acc.*) water, others to go forward that they might choose a place for the camp. 5. Well knowing therefore that you have few arms we have come bringing horses and swords (ξίφος *neut.*). 6. After crossing the river they went up into the mountains. 7. The passers-by (πάρειμι) saw that we were being insulted by the barbarians; therefore they came forward willingly to save the women and children. 8. But calling together all whom he knew to be friendly (φίλιος) to himself he spoke as follows: "Since so great a disaster has fallen upon us (ἐπι-πίπτω *with dat.*) it is impossible any-longer to go against the enemy; let us therefore retreat as quickly as possible." 9. Whenever he comes to our city we receive him (as) a guest. 10. Do not go out of the city but stay at home to-day. 11. Let nothing prevent you from going to help those who are being enslaved by this tyrant.

INDIRECT STATEMENT

Pres. Act.	*Fut. Act.*	*Aor. Act.*	*Perf. Act.*	*Perf. Mid. and Pass.*	*Aor. Pass.*
λέγω	λέξω	ἔλεξα		λέλεγμαι	ἐλέχθην
φημί	φήσω	ἔφην		-είλεγμαι	-ελέγην
ἀγορεύω	ἐρῶ	εἶπον	εἴρηκα	εἴρημαι	ἐρρήθην

φημί, φῄς, φησί(ν), φαμέν, φατέ, φασί(ν).

ἔφην, ἔφησθα, ἔφη, ἔφαμεν, ἔφατε, ἔφασαν.

Subj. φῶ. *Optat.* φαίην. *Infin.* φάναι. *Partic.* φάσκων.

φημί, οἴομαι 'think' are used (1) with actual quotations, (2) with infin. (acc. *or* nom.).

The imperf. ἔφην is used as an aorist. ἔλεγον is the imperfect. The meaning of φημί is 'I assert,' stronger than λέγω.

ἔφασκον imperf. means 'they alleged' (implying falsehood).

λέγω in compounds has the meaning 'gather,' *e.g.* συλλέγω 'collect,' ἐκλέγω 'pick out.'

ἀγορεύω 'make a public speech' is used for the pres. and imperf. of compounds. προαγορεύω 'proclaim,' ἀπαγορεύω 'forbid, renounce.'

ἐρῶ is contracted for ἐρέω (p. 54).

εἶπας for εἶπες, εἴπατε for εἴπετε, εἰπάτω for εἰπέτω are usual.

λέγω and other verbs of saying and thinking take ὅτι and indic. after a primary tense, optative after a historic tense, but the indic. is often retained after a historic tense. The *tense* is never altered. ὡς may be used for ὅτι if any doubt is cast upon the statement.

INDIRECT QUESTION

The same principles apply to indirect questions. An indirect question can only be in the subjunctive if the original question was in the subj. (p. 61).

Indirect questions can only follow verbs of saying, thinking and asking; or equivalent phrases.

διαλέγομαι 'converse' (dialogue).

Pres. ἐρωτάω *Fut.* ἐρωτήσω *Aor.* ἠρώτησα 'I ask' (a question).
ἐρήσομαι ἠρόμην.

For the other sort of asking = begging, αἰτέω and its compounds must be used.

πότερον, πότερα...ἤ 'whether...or' (double question).

εἰ may be used for 'if' = 'whether,' as in English; 'or not' = ἢ οὐ *or* ἢ μή.

ἀπορέω to be at a loss.
ὁ αὐτόμολος deserter.
μῶρος -α -ον foolish, stupid.
σιωπάω, σιγάω be silent.
μάχομαι fight, *fut.* μαχοῦμαι (*p.* 48), *aor.* ἐμαχεσάμην.
τυφλός -ή -όν blind.

Ex. 74. 1. ἤρετο μὲν ὁ κῆρυξ, Τίς ἀγορεύειν βούλεται; παρῄει δ' οὐδείς.
2. ὁ κῆρυξ ἤρετο ὅστις ἀγορεύειν βούλοιτο.
3. ἀλλ' εἰπέ μοι, τίς ἔλεξε τάδε· Ὅσα εἴρηκα εἴρηκα;
4. ὁ δὲ Κλέαρχος ἤρετο τὸν Κῦρον· Οἴει γάρ σοι μαχεῖσθαι, ὦ Κῦρε, τὸν ἀδελφόν; Νὴ τοὺς θεούς, ἔφη ὁ Κῦρος, εἴπερ γε Δαρείου καὶ Παρυσάτιδός ἐστι παῖς, ἐμὸς δὲ ἀδελφός, οὐκ ἀμαχεὶ τὴν ἐκείνου ἀρχὴν λήψομαι. ἐν ᾧ δὲ οὕτως διελέγοντο ἧκέ τις ἀγγέλλων ὅτι ὁ βασιλεὺς πάρεστι (*or* παρείη) ἔχων πολλὴν δύναμιν, εἰς μάχην παρεσκευασμένος.
5. Ὤετο μὲν γὰρ εἰδέναι πολλά, ἔφη ὁ Σωκράτης, τῷ δὲ ὄντι ᾔδει οὐδέν.
6. οἱ μὲν δὴ ἄγγελοι ταῦτα ἀκούσαντες ᾤχοντο. ἐν τούτῳ δὲ ἧκε Χρυσάντας ὁ Πέρσης καὶ ἄλλοι, αὐτομόλους ἄγοντες· καὶ ὁ βασιλεὺς ἠρώτα τοὺς αὐτομόλους τί ποιοῖεν οἱ πολέμιοι· οἱ δ' ἔλεγον ὅτι ἐξίοιέν τ' ἤδη σὺν (with) τοῖς ὅπλοις καὶ παρατάττοι αὐτοὺς αὐτὸς ὁ σατράπης (governor).
7. ἐσίγα οὖν ἀπορῶν τί λέγοι (*or* λέγῃ).
8. περὶ μὲν οὖν τούτων τοσαῦτά μοι εἰρήσθω, νῦν δὲ δηλώσω τὰ μετὰ ταῦτα γενόμενα.
9. ὁ δὲ ἄγγελος ἠρώτησε, Σπονδὰς ἢ πόλεμον ἀπ-αγγελῶ; Κλέαρχος δὲ αὖθις ταὐτὰ ἀπεκρίνατο, Σπονδαὶ μὲν ἡμῖν ἔσονται μένουσιν, ἀπιοῦσι δὲ ἢ προϊοῦσι πόλεμος. ὅτι δὲ ποιήσοι οὐ δι-εσήμηνεν (*p.* 54).

Ex. 75. 1. I do not know whether we ought to speak or to be silent. 2. I for my part think that it is better to fight at once that the enemy may not become stronger. 3. But Cleon said that not he himself but Nikias was general. 4. You think you are very wise, but we know you are very foolish. 5. The boy however answered that he honoured (*pres. opt.*) his father and his mother, and that he would obey them rather than the general. 6. Surely you will not be so foolish as to say that you have done us no harm since we ourselves saw you doing these very things? 7. Many things have been said concerning Homer: first, that he was blind; secondly, that he did not write the poems which are said to have been written by him; thirdly, that such a man did not even (οὐδέ) exist ever. 8. When these things had been said, some one asked why (διὰ τί) it was impossible to cross those mountains which Hannibal (Ἀννίβας) once crossed with (ἄγων) a great army and many horses and elephants and that too (καὶ ταῦτα) in winter.

VERBS OF FEARING

	Pres.	*Fut.*	*Perf.*	*Aor.*
Mid.& Pass.	φοβέομαι	φοβήσομαι	πεφόβημαι	*Pass.* ἐφοβήθην
Act.	(δείδω)	δείσω	*Partic.* δεδιώς.	ἔδεισα

δέδοικα (indic. only) perfect with present meaning 'I am apprehensive lest....'

These verbs may be followed by (1) an Acc., (2) an Infinitive, (3) μή and Subj. after a Primary Tense, Optative after Historic Tense to express fear for the future, (4) Indicative to express fear for present or past. The Negative is μὴ οὐ; μή = lest, *or* that.

(1) *οἱ ἐφ' ἵππων φοβοῦνται, οὐχ ἡμᾶς μόνον, ἀλλὰ τὸ καταπεσεῖν.*

(2) μὴ φοβηθῇς εἰσελθεῖν· οὐ γὰρ ἔτι κίνδυνός ('danger') ἐστι.

(3) *ἔδεισαν οἱ Ἕλληνες μὴ κυκλῶνται* (*κυκλόω* 'surround') *or κυκλοῖντο.*

(4) *δέδοικα μὴ ὁ στρατὸς ἡμῶν νενίκηται.*

VERBS OF PRECAUTION

ἐπιμελέομαι 'take pains,' φυλάττομαι 'I am on my guard.'

εὐλαβέομαι 'beware,' σκοπέω 'look out, see to it that' (σκέψομαι, ἐσκεψάμην, ἔσκεμμαι).

ὁράω, πράττω, παρασκευάζω are often used in this way.

They are followed by (1) ὅπως with a purpose clause (p. 62 (1)) *or* (2) ὅπως with the Fut. Indic. Negative always μή.

When both verbs are in the 2nd person the precaution-verb is often omitted.

(σκοπεῖτε) ὅπως ἔσεσθε ἄνδρες ἄξιοι ἐλευθερίας
See that you show yourselves men worthy of freedom.

εὐλαβεῖσθαι οὖν δεῖ ὅπως μὴ οἱ πολέμιοι προσίωσιν
We must take care that the enemy do not approach.

ὁ στρατηγὸς ἐπιμελήσεται ὅπως τὰ ἐπιτήδεια ἕξομεν
The general will see to it that we have provisions.

SOME IRREGULAR VERBS

Pres.		*Fut.*	*Aor.*	*Perf.*
πάσχω	experience, suffer	πείσομαι	ἔπαθον	πέπονθα
ἐσθίω	eat	ἔδομαι	ἔφαγον	(ἐδήδοκα)
ὁράω	see	ὄψομαι	εἶδον	ἑόρακα

Perf. Pass. ὦμμαι. *Aor. Pass.* ὤφθην. *Imperf.* ἑώρων.

Distinguish: ἐλήφθην 'I was taken'; ἐλείφθην 'I was left.'

ὁ ἡ αἴλουρος cat.
ὁ φόβος fear.
ὁ ὕπνος sleep.
καθεύδω sleep, *fut.* καθευδήσω.
τὸ τέκνον child.
ὁ δεσμός chain, *pl.* prison.
ἔνιοι some.
ὁ μῦς *gen.* μυός mouse.
τὸ δέος apprehension, fear.
ὁ ἡ φυγάς φυγάδ- exile, fugitive.
ναυαγέω suffer shipwreck.
μηχανάομαι contrive.
διακεῖμαι to be situated.
ἐνίοτε sometimes.

Ex. 76. νῦν δὲ φοβούμεθα μὴ ταῦτα τὰ ἀπ-ηγγελμένα οὐκ ἔστιν ἀληθῆ.
2. Εὖ ἴσθι, ὦ φίλε, ἔφη, ὅτι οὔτ' ἐν πολέμῳ οὔτ' ἐν δίκῃ (law) οὐδένα δεῖ τοῦτο μηχανᾶσθαι ὅπως ἀποφεύξεται, πᾶν ποιῶν, θάνατον.
3. οὐ δέος ἐστὶ μὴ ἴδωσιν ἡμᾶς.
4. ἐφοβήθησαν γὰρ μὴ αὐτοὶ τὰ αὐτὰ πάθοιεν.
5. ὁ αἴλουρος ἐφοβεῖτο μὴ οὐ συλλάβῃ τὸν μῦν.
6. Καὶ σύ γε, ἔφη, οἶσθα, ὅτι ἀληθῆ λέγω· ἐπίστασαι γάρ, ὅτι οἱ μὲν φοβούμενοι μὴ φύγωσι πατρίδα, καὶ οἱ μέλλοντες (about to) μαχεῖσθαι, δεδιότες μὴ νικηθῶσιν, ἀθύμως διακεῖνται καὶ οἱ πλέοντες μὴ ναυαγήσωσιν, καὶ οἱ δουλείαν καὶ δεσμοὺς φοβούμενοι, οὗτοι μὲν οὔτε σῖτον οὔθ' ὕπνον λαγχάνειν (obtain) δύνανται διὰ τὸν φόβον· οἱ δὲ ἤδη φυγάδες, ἤδη δὲ νενικημένοι, ἤδη δὲ δουλεύοντες, ἐνίοτε δύνανται καὶ μᾶλλον τῶν εὐδαιμόνων ἐσθίειν τε καὶ καθεύδειν. καὶ δὴ καὶ ἔνιοι φοβούμενοι μὴ ληφθέντες ἀποθάνωσιν, ὑπὸ τοῦ φόβου προαποθνήσκουσιν. οὕτω πάντων τῶν δεινῶν ὁ φόβος μάλιστα καταπλήττει (confounds) τὰς ψυχάς. τὸν δ' ἐμὸν πατέρα, ἔφη, νῦν πῶς δοκεῖς διακεῖσθαι τὴν ψυχήν, ὃς οὐ μόνον περὶ ἑαυτοῦ, ἀλλὰ καὶ περὶ ἐμοῦ τοῦ ἑαυτοῦ παιδὸς καὶ περὶ γυναικὸς καὶ περὶ πάντων τῶν τέκνων δουλείας φοβεῖται;

Ex. 77. 1. It will not be difficult to contrive that he shall escape from prison.
2. We were afraid that the enemy would burn the city which we abandoned.
3. I am afraid that you have greatly injured both yourself and us.
4. We have suffered much, o soldiers, for our country and I fear that we shall suffer even more terrible things in the present war.
5. So great a panic fell upon (ἐμ-πίπτω) the citizens that each fled to his own house, fearing that he might be taken-prisoner.
6. Of all the cities which (*p.* 20 (2)) I have seen this seems to me (the) ugliest.
7. He died in his sleep (*pres. part.*).
8. (See) that you do not tell anyone about this.
9. They therefore were unable to sleep that night, fearing that they themselves would suffer the same things which they had done to others.
10. Are you afraid to speak?

δί-δω-μι 'I give.' *Verb Stem* δο-. *Pres. Stem* δι-δο-

	Pres. Indic.	*Pres. Imper.*	*Aor. Indic.*	*Aor. Imper.*	*Imperf. Indic.*
Sing.	δίδωμι	—	ἔδωκα	—	ἐδίδουν
	δίδως	δίδου	ἔδωκας	δός	ἐδίδους
	δίδωσι(ν)	διδότω	ἔδωκε(ν)	δότω	ἐδίδου
Plur.	δίδομεν	—	ἔδομεν	—	ἐδίδομεν
	δίδοτε	δίδοτε	ἔδοτε	δότε	ἐδίδοτε
	διδόασι(ν)	διδόντων	ἔδοσαν	δόντων	ἐδίδοσαν
			(ἔδωκαν)		(I was giving *or* I offered)
Subj.	διδῶ (*cp.* γνῶ, p. 68)		δῶ		
Opt.	διδοίην		δοίην		
Infin.	διδόναι		δοῦναι		*Perf.*
Part.	διδούς (*cp.* γνούς, p. 68)		δούς		δέδωκα
Pass.			ἐδόθην		δέδομαι

Pres. Mid. and Pass. δίδομαι. δίδοσαι etc. *Imperf.* ἐδιδόμην ἐδίδοσο etc.

Aor. Mid. ἐδόμην, ἔδου etc. *Infin.* δόσθαι. *Part.* δόμενος.

CONDITIONAL CLAUSES ('if, unless'). I

(*a*) Present or Past Time. Fulfilled condition (the speaker assumes the fulfilment of the condition): εἰ with the Indicative (Prŏtasis); main verb (Apŏdŏsis) Indicative or Imperative, *e.g.*

If they have betrayed the city, they are worthy of death
εἰ προδεδώκασι τὴν πόλιν, ἄξιοί εἰσι θανάτου.

Whether you are rich or poor (or not) you may not do this
εἴτε πλούσιος εἶ εἴτε πένης (εἴτε μή), τοῦτο οὐκ ἔξεστί σοι.

(*b*) Present or Past Time. Unfulfilled condition (the speaker states what would be happening or would have happened, if things had been otherwise): εἰ with the Imperfect or Aorist Indic.; main clause with ἄν.

The Imperfect means an action in present time or a continued action in past time ('had been doing'); the Aorist as usual an action occurring in the past ('had done').

If they had betrayed the city, they would have been slain.
εἰ προέδοσαν (προὔδοσαν) τὴν πόλιν, ἀπέθανον ἄν.

If you had been well educated, you would not (now) be ignorant of this
εἰ εὖ ἐπαιδεύθης, οὐκ ἂν ἠγνόεις ταῦτα.

προδίδωμι betray. παραδίδωμι hand over, surrender.
ἀποδίδωμι give back. ἀποδίδομαι (*mid.*) sell.

These are followed by acc. (and dative of the recipient), but μεταδίδωμι 'give a share' will have gen. of the thing divided.

ἐνδίδωμι surrender, give in (*intr.*).

τὸ μέλι μέλιτ- honey. πάνυ very, altogether.
γεύομαι taste (*gen.*). καταφρονέω despise (*gen.*).

Ex. 78. 1. εἴ με ἐκέλευσας ταῦτα ποιεῖν, ἡδέως ἂν ἐποίησα.
2. οὐκ ἂν αὖθίς σε ἠρώτων εἰ πρότερον ἀπεκρίνω.
3. εἰ μὲν γὰρ ἡμεῖς αὐτοὶ πρός τε τὴν πόλιν ἐλθόντες ἐμαχόμεθα καὶ τὴν γῆν ἐδῃοῦμεν (plunder) ὡς πολέμιοι, ἀδικοῦμεν· εἰ δὲ ὑμῶν οἱ πρῶτοι ἐπεκαλέσαντο (invite) ἡμᾶς, τί ἀδικοῦμεν;
4. εἰ δ' ἦσαν ἄνδρες ἀγαθοί, οὐκ ἂν ταῦτα ἔπασχον.
5. εἰ ἅπαντα ᾔδησθα, οὐκ ἂν ἐμοῦ κατέγνως.
6. εἴ τις ψεύδεται, τί ἄλλο ἢ ἀνθρώπους φοβούμενος θεοῦ καταφρονεῖ;
7. Κυμαῖός τις μέλι ἐπώλει· ἐλθόντος δέ τινος καὶ γευσαμένου καὶ εἰπόντος, Πάνυ καλόν ἐστι· Εἰ μὴ γάρ, ἔφη, μῦς ἐνέπεσεν εἰς αὐτό, οὐκ ἂν ἐπώλουν.

In reported speech (*a*) the apodosis may be put in the infinitive retaining ἄν (neg. οὐ).

A participle often takes the place of the protasis (neg. μή).

ἔφασαν αὐτὸν σωθῆναι ἄν, εἰ νεῖν ἠπίστατο

They said he would have been saved if he had known how to swim.

Participle: ἐσώθη ἂν νεῖν ἐπιστάμενος.

Ex. 79. 1. If we had given them money, they would have released us.
2. If you had been persuaded by my words, you would not now be suffering such (things).
3. If you despise the enemy, you are very foolish.
4. He however says that he would have come (*aor. infin.*) if we had invited him.
5. What would you have done if you had found a burglar in your house at night?
6. If he says that we gave him money, he is lying.
7. Those who betrayed the city to the enemy would have been put to death if they had been caught.
8. He said he would have fallen, if you had not saved him.
9. If everything has been prepared, let us march.
10. If Demosthenes were present to-day, he would now be speaking in our defence (πρό *with gen.*).
11. When he tasted the honey he said it was very good.
12. If that is so, why have you not given him a share of the money?

CONDITIONAL CLAUSES. II

Future Time: (1) The ordinary construction is *ἐάν* (*ἤν or ἄν*) with the Subj. (Aorist for a single event); apodosis in the Fut. Indic.: 'they will do it, if they can' *τοῦτο ποιήσουσιν ἢν δύνωνται. ἐάν τι εὕρῃς, δός μοι* 'if you find anything, give it to me.'

(2) *εἰ* with the Fut. Indic. is especially used for threats and warnings: *εἰ μὴ καθέξεις γλῶτταν ἔσται σοι κακά* 'if you don't restrain your tongue, it will be bad for you.' (Note the present tense in the English.)

(3) For a vaguer indistinct future the Optative is used for both, the main clause having *ἄν*: 'they would help us if they could' *ὠφελοῖεν ἂν ἡμᾶς, εἰ δύναιντο.* The if-clause is often understood (p. 66 (2)).

In all these the apodosis may be a participle or an infinitive if the construction requires it; *ἄν* of course being retained. *οἶδα ὑμᾶς, ἢν τοῦτο ποιήσητε, εὖ πράξοντας* 'I know you will fare well if you do this.'

N.B. *ἄν* in Greek = 'would' in English = Subj. in Latin.

(4) The phrase 'on condition that' is expressed by *ἐφ' ᾧ or ἐφ' ᾧτε* with (*a*) the Fut. Indic., (*b*) an Infinitive or Acc. and Infinitive. The Negative is *μή*. *ὥστε* (*μή*) is sometimes used but only with the Infinitive.

(*a*) *συνέβησαν ἐφ' ᾧτε ἐξίασιν ἐκ Πελοποννήσου ὑπόσπονδοι καὶ μηδέποτε ἐπιβήσονται αὐτῆς.* (*ἐξίασιν* might be *ἐξιέναι* and *ἐπιβήσονται ἐπιβήσεσθαι*, but *ὑπόσπονδοι* would still be nom. case.) 'They agreed to leave the Peloponnesus under a truce and never to set foot on it again.'

(*b*) *ἔπειτα δ' ὕστερον καὶ πρὸς τοὺς ἄλλους ἅπαντας τοὺς μετὰ Δημοσθένους ὁμολογία γίγνεται ὥστε ὅπλα τε παραδοῦναι καὶ μὴ ἀποθανεῖν μηδένα μήτε βιαίως, μήτε δεσμοῖς, μήτε σίτου ἐνδείᾳ* 'Then afterwards an agreement was made also with all the rest that were with Demosthenes that they should give up their arms and no one should be put to death either by violence or by imprisonment or by lack of food.'

N.B. *εἴ τις* (*καὶ*) *ἄλλος* = 'si quis alius.'

ὁ Θεμιστοκλῆς, εἴ τις καὶ ἄλλος, εὐηργέτησε τὴν πατρίδα
No one surpassed Themistocles as a benefactor of his country.

Note: sentences beginning with 'if ever' (really temporal): Primary (*ἐάν* and Subj.), Historic (*εἰ* and Optative). See pages 64 (2) and 66 (3).

δάκνω bite, δήξομαι, ἔδακον, δέδηχα, δέδηγμαι, ἐδήχθην.
δειλός -ή -όν cowardly. νέος, νέα, νέον, new, fresh, young.
ἀκολουθέω follow (*dat.*). ὥσπερ like, as, just as.
πρόσθεν, ἔμπροσθεν *adverbs and preps. with gen.* in front.
ὄπισθεν, ὀπίσω behind (*of time*, hereafter).
δίκην λαβεῖν to exact the penalty. δίκην δοῦναι to pay the penalty.

Ex. 80. 1. τούτων οὕτως ἐχόντων ἀπο-λύσομέν σε, ἐπὶ τούτῳ μέντοι, ἐφ' ᾧτε ἀποθανεῖ ἐάν ποτε ἁλῷς ἐν τῇ ἡμετέρᾳ (χώρᾳ).

2. δειλότατος ἂν εἴη ὁ ναύκληρος (ship-owner) εἰ ἀπολίποι τὴν ναῦν μὴ σωθέντων πάντων τῶν ναυτῶν.

3. εἴ τι κλέψεις τῶν σοι παραδοθέντων, δίκην δώσεις κατὰ τὸν νόμον.

4. οὐκ ἔφασαν ἰέναι οἱ στρατιῶται, ἐὰν μή τις σφίσι χρήματα διδῷ.

5. ἀλλ' ἢν φοβήθῃς μὴ ἁλῷς ὑπὸ τῶν λῃστῶν (bandits) συνεψόμεθά σοι οἴκαδε.

6. σύ, εἴ τις ἄλλος, οἶμαι, ὠφελοίης ἄν, εἴ τι δέοι.

7. εἰ μὲν οὖν ἄλλος τις βέλτιον ὁρᾷ (*Indic. or Subj.?*), ἄλλως ἐχέτω, εἰ δὲ μή, Χειρίσοφος μὲν ἡγείσθω, ἐπειδὴ καὶ Λακεδαιμόνιός ἐστι· ὀπισθο-φυλακῶμεν δ' ἡμεῖς οἱ νεώτατοι. οὐ γὰρ θαυμάζοιμ' ἄν, εἰ οἱ πολέμιοι, ὥσπερ οἱ δειλοὶ κύνες τοὺς μὲν παριόντας (*which* πάρειμι*?*) διώκουσί τε καὶ δάκνουσιν, ἢν δύνωνται, τοὺς δὲ διώκοντας φεύγουσιν, εἰ καὶ αὐτοὶ ἡμῖν ἀπιοῦσιν ἐπ-ακολουθοῖεν.

8. ἀποθάνοιμεν ἂν εὐθύς, εἴ ποθ', ὃ μὴ γένοιτο, οἱ πολέμιοι ἡμᾶς ἴδοιεν.

9. ἐὰν ταῦτα ἀληθῆ ᾖ, οὐδέποτε αὖθις πιστεύσω ἐκείνῳ.

Ex. 81. 1. I would never forgive you if you were to abandon us under these circumstances. 2. The dog will follow you if you speak (προσεῖπον) pleasantly to it. 3. If he has not arrived, I fear something terrible has happened. 4. If they do anything contrary to the law, they shall pay the penalty. 5. If they had made peace (εἰρήνη) on those terms, the Persians would have enslaved the whole of Greece. 6. If you do what you promised all will be well, but if not, we shall prevent you from returning home. 7. He refused to help us unless we gave him more money. 8. I think they would all have been saved if they had known how to swim. 9. You would easily defeat them, if you had more horsemen. 10. They therefore made (*mid.*) a treaty on condition that both armies (ἀμφότεροι) should depart and not (μηδέ) injure (βλάπτω) each other's land. 11. What would you do if you found a thief in your house at night?

PARTICIPLES: FURTHER USES (p. 44)

(1) *μεταξύ* 'in the midst,' *ἅμα* 'together,' *εὐθύς* 'immediately,' are often used for greater precision with a participle: *ἅμα φεύγοντες τὰ ὅπλα ἀπέβαλλον* 'while fleeing they threw away their arms'; *μεταξὺ δειπνοῦντες συνεβουλεύοντο* 'while dining they made their plans'; *εὐθὺς ἰδὼν τὸν ἵππον ἐβουλήθη ἔχειν* 'as soon as he saw the horse he wanted to have it.'

μεταξύ (*gen.*), *ἅμα* (*dat.*), are also used as prepositions.

(2) Verbs of beginning, continuing, and ceasing, have usually a participle following: *ἄρχομαι, διατελέω, παύομαι.* So also *τυγχάνω* 'I happen to be,' *χαίρω, ἥδομαι* 'take delight in,' *λανθάνω* 'escape notice,' *φθάνω* 'anticipate' (= do something before some other event), *περιοράω* 'overlook.'

(3) *φαίνομαι* and *αἰσχύνομαι* are used with participle or infinitive, the meaning varying accordingly.

φαίνεται ὀργιζόμενός μοι 'he is evidently becoming angry with me.'

φαίνεται ὀργίζεσθαι 'he seems to be angry (but is not).'

αἰσχύνομαι λέγων 'I say with shame.'

αἰσχύνομαι λέγειν 'I am ashamed to say (and do not say).'

ἔτυχον παρόντες 'they happened to be present.'

ἔλαθεν ἡμᾶς φυγών or
λαθὼν ἡμᾶς ἔφυγεν } 'he escaped without our knowing it.'

ἔφθασεν ἡμᾶς ἀφικόμενος or
φθάσας ἡμᾶς ἀφίκετο } 'he arrived before we did.'

οἱ νέοι χαίρουσι ζῶντες 'the young enjoy living.'

(4) *δῆλός εἰμι*
φανερός εἰμι } are used like *φαίνομαι* with a participle and with the same meaning.

φαίνομαι	*φανοῦμαι*	*ἐφάνην*
αἰσχύνομαι	*αἰσχυνοῦμαι*	*ᾐσχύνθην*
τυγχάνω	*τεύξομαι*	*ἔτυχον*
ὀργίζομαι	*ὀργισθήσομαι*	*ὠργίσθην*
ἥδομαι	*ἡσθήσομαι*	*ἥσθην*
φθάνω	*φθήσομαι*	*ἔφθασα* and *ἔφθην*
λανθάνω	*λήσω*	*ἔλαθον, perf. λέληθα*
θέω, τρέχω	*δραμοῦμαι*	*ἔδραμον* I run

ὁ πόνος labour, trouble. ἔστε until.

Ex. 82. 1. δῆλος εἶ οὐδὲν παρὰ τοῦ διδασκάλου μεμαθηκώς, καίπερ πολλὰ δοὺς αὐτῷ χρήματα.

2. ἔτυχον καθεύδοντες ἐν τῇ ἀγορᾷ ἐννέα στρατιῶται.

3. χαίρουσι τιμώμενοι ὑπ' ἀνθρώπων καὶ οἱ θεοί.

4. *Goodbye to all that!*

ἐκ δὲ τούτου συν-ελθόντες συν-εβουλεύοντο περὶ τῆς λοιπῆς πορείας· ἀνέστη (stood up) δὲ πρῶτος Ἀντιλέων καὶ ἔλεξεν ὧδε· Ἐγὼ μὲν τοίνυν (therefore), ἔφη, ὦ ἄνδρες, ἀπείρηκα (am tired of) ἤδη συσκευαζόμενος καὶ πορευόμενος, καὶ βαδίζων καὶ τρέχων καὶ τὰ ὅπλα φέρων καὶ ἐν τάξει ἰὼν καὶ φυλακὰς φυλάττων καὶ μαχόμενος, ἐπιθυμῶ δε ἤδη, παυσάμενος τούτων τῶν πόνων, ἐπεὶ θάλατταν ἔχομεν, πλεῖν τὸ λοιπὸν καί, ὥσπερ Ὀδυσσεύς, καθεύδων ἀφικέσθαι εἰς τὴν Ἑλλάδα. καὶ πάντες ταὐτὰ ἔλεγον οἱ παρόντες. ἔπειτα δὲ ὁ Χειρίσοφος ἀνα-στὰς εἶπεν ὧδε· Φίλος μοί ἐστιν, ὦ ἄνδρες, Ἀναξίβιος, ναυαρχῶν δὲ τυγχάνει. ἢν οὖν πέμψητέ με, οἴομαι ἂν ἐλθεῖν καὶ τριήρεις ἔχων καὶ πλοῖα τὰ ἡμᾶς ἄξοντα· ὑμεῖς δὲ εἴ-περ πλεῖν βούλεσθε, περι-μένετε ἔστε ἂν ἐγὼ ἔλθω· ἥξω δὲ ταχέως. οἱ δὲ στρατιῶται ἀκούσαντες ταῦτα ἥσθησάν τε καὶ ἐψηφίσαντο πλεῖν ὡς τάχιστα.

5. μὴ περι-ίδωμεν τὴν πόλιν διαφθειρομένην.

6. τοσοῦτο ἐφθάσαμεν ἀφικόμενοι, ὥστε ἡ συμφορὰ οὐκ ἤγγελτο.

7. κλέπται ἐφάνησαν ὄντες.

8. βουλοίμην ἂν λαθεῖν αὐτὸν ἀπελθών.

9. φαίνεται εἶναι σόφος.

Ex. 83. 1. If you can enter the enemy's camp (στρατόπεδον) without the knowledge of the sentinels, you will learn what they are planning (βουλεύομαι). 2. Directly the thief saw me he shut the door and ran away. 3. We all rejoiced when we heard that you had ceased at last to be angry with your father. 4. He was plainly ashamed to confess what he had done. 5. We shall never cease to thank you (χάριν οἶδα *with dat.*) for what (ἀνθ' ὧν) you did on our behalf. 6. If I had happened to be present, I should have prevented this. 7. They marched as quickly as possible in order to reach the city first. 8. Do not look on at the execution of this brave man. 9. It is clear that they are afraid that they may be noticed (οὐ λαθόντες) crossing the plain; otherwise (εἰ δὲ μή) they would not have waited until night came. 10. Cease (παῦε *not* παύου) to despise those who are not rich. 11. Were you not ashamed to say such things? 12. If you were to arrive at daybreak (ἅμα), you would take the city unawares (ἀπροσδόκητος). 78 (3).

SOME IRREGULAR NOUNS

ἡ θρίξ	hair	τριχ-,	*dat. pl.* θριξί(ν).
τὸ φῶς	light	φωτ-,	*no plural.*

τὸ κρέας 'meat' like κέρας, but has contracted forms only.

τὸ φρέαρ	well	φρεατ-	*dat. pl.* -ασι(ν).
τὸ ἧπαρ	liver	ἡπατ-	*dat. pl.* -ασι(ν).
τὸ γόνυ	knee	γονατ-	*dat. pl.* -ασι(ν).
τὸ δόρυ	spear	δορατ-	*dat. pl.* -ασι(ν).
τὸ ὄναρ	dream	ὀνειρατ-	*dat. pl.* -ασι(ν).
τὸ οὖς	ear	ὠτ-	
τὸ γάλα	milk	γαλακτ-	
τὸ πῦρ	fire	πυρ- (*pl. as if from* πυρόν 'watchfires').	
ὁ ἀστήρ	star	ἀστερ-, *dat. pl.* ἀστράσι(ν).	
ἡ ἠχώ	echo	*acc.* ἠχώ, *gen.* ἠχοῦς, *dat.* ἠχοῖ.	
ἡ ἕως	dawn, east	*acc.* ἕω, *gen.* ἕω, *dat.* ἕῳ.	
ὁ Ζεύς	Zeus	*acc.* Δία, *gen.* Διός, *dat.* Διΐ, *voc.* Ζεῦ.	
ὁ ἡ μάρτυς	witness	μαρτυρ-, *dat. pl.* μάρτυσι(ν).	

τὸ κέρας 'horn' (or 'wing' of an army), κερατ- but the τ is often dropped and contraction made: gen. κέρατος *or* κέρως.

ATTIC DECLENSION

Some nouns in the 2nd decl. have a long ω throughout.

	Sing.		Plur.
Nom.	ὁ νεώς	temple	νεῴ
Acc.	νεών		νεώς
Gen.	νεώ		νεών
Dat.	νεῴ		νεῴς

so also ὁ λεώς 'people,' ὁ καλώς 'cable,' and the adjectives ἵλεως 'propitious,' πλέως 'full' (fem. πλέα).

The forms of the following verbs are easily confused.

τρέφω	τρέπω	στρέφω
θρέψω	τρέψω	στρέψω
ἔθρεψα	ἔτρεψα, ἔτραπον	ἔστρεψα
τέτροφα	τέτροφα	ἔστροφα
τέθραμμαι	τέτραμμαι	ἔστραμμαι
ἐτράφην	ἐτράπην, ἐτρέφθην	ἐστράφην
nourish	turn	turn about

τρέπω has two aorists middle, ἐτρεψάμην, ἐτραπόμην, and is thus the only verb which has all six possible aorists.

δεξιός -ά -όν right (hand).
ἡ χελιδών χελιδον- swallow.
ἔνθα there, then, where.
ἀριστερός, εὐώνυμος -ον left (hand).
τὸ στόμα mouth.
ἀμφί about, around.

Ex. 84. 1. μία χελιδὼν οὐκ ἔαρ ποιεῖ.
2. διὰ τοῦτο ὦτα μὲν δύο ἔχομεν, στόμα δὲ ἕν, ἵνα πλείω μὲν ἀκούωμεν, ἧττω δὲ λέγωμεν.
3. ἔνθα δὴ οἱ μὲν βάρβαροι στραφέντες ἔφευγον ᾗ ἕκαστος ἐδύνατο· οἱ δ' ἀμφὶ Τισσαφέρνην ἀποτραπόμενοι ἄλλην ὁδὸν ᾤχοντο.
4. Κλέαρχος μὲν τὰ δεξιὰ τοῦ κέρατος, Μένων δὲ τὸ εὐώνυμον κέρας ἔσχε τοῦ Ἑλληνικοῦ.
5. ὅρα μὴ καταπέσῃς εἰς τὸ φρέαρ.
6. ὦ Ζεῦ, μάρτυρα σὲ ποιῶ ὧν πάσχω.
7. ὁ παῖς μελαίνας ἔχει τὰς τρίχας.
8. ἐν ταύταις ταῖς κώμαις εὗρον τοὺς τῶν βαρβάρων παῖδας μάλα εὖ τεθραμμένους γάλακτι καὶ κρέασι τῶν βοῶν.
9. ἀνάγνωθί μοι τοὺς τοῦ μάρτυρος λόγους.
10. καθ' Ὅμηρον, εἰσὶ δύο πύλαι τοῦ ὕπνου· ἐκ μὲν τῆς κερατίνης πύλης πέμπονται παρὰ Διὸς τὰ ἀληθῆ ὀνείρατα, ἐκ δὲ τῆς ἄλλης, τῆς ἐλεφαντίνης, τὰ ψευδῆ.
11. φιλόσοφός τις ἔφη ἐκ τοῦ πυρὸς γενέσθαι πάντα· ἄνευ γὰρ τούτου οὔτε τὸν ἥλιον οὔτε τὴν σελήνην οὔτε τοὺς ἀστέρας τὸ φῶς δύνασθαι παρέχειν· καὶ μὴν καὶ οὐδὲν δυνατὸν εἶναι ζῆν, θέρους ἀπόντος.

12. *Olympicus the prophet answers wisely.*

εἰς Ῥόδον εἰ πλεῖν δεῖ, τις Ὀλυμπικὸν ἦλθεν ἐρωτῶν
τὸν μάντιν, καὶ πῶς πλεύσεται ἀσφαλέως.
χὠ (καὶ ὁ) μάντις Πρῶτον μέν, ἔφη, καινὴν ἔχε τὴν ναῦν,
καὶ μὴ χειμῶνος, τοῦ δὲ θέρους ἀνάγου (set sail).
ταῦτα γὰρ ἢν ποιῇς, ἥξεις κἀκεῖσε καὶ ὧδε (there and back),
ἢν μὴ πειρατὴς ἐν πελάγει σε λάβῃ.

Ex. 85. 1. At dawn (ἅμα *with dat.*) the light of the stars is no longer seen. 2. He came with (ἔχων) a spear in each hand (ἑκάτερος). 3. The left wing of the enemy turned (*aor. part. mid.*) and fled. 4. This is the same child, whom I brought up with my own hands. 5. If you ask the echo (a question) it will give you back your own words again. 6. By Zeus, said he, I will not look on while my country is being enslaved. 7. They happened to have lighted fires in the market-place. 8. If you were not to honour the gods by sacrificing (θύω) in their temples, they would not be favourable. 9. Turning (*aor. part. mid. or pass.*) to his brother he asked him what was the name of that star; and he (*p.* 18 (5)) replied that it was called the Lion. 10. He used to relate (ἐξηγέομαι) his dreams to everybody whether they wished to hear him or not.

VERBAL ADJECTIVES

These are formed from the verb stem with **-σιμος**, **-τος**, -τέος. χράομαι 'I use,' χρήσιμος 'useful,' χρηστός 'excellent.' These two forms merely express the quality derived from the verb's meaning.

The negative form of adjectives in -τος is very common owing to the liking of the Greek genius for negative expressions such as οὐκ ἀπαίδευτος 'not uneducated.' The negative is α- before a consonant, αν- before a vowel. εὐ- 'well' and δυσ- 'scarcely' are often prefixed.

Before -τος and -τέος the consonants change as on p. 40. κ, γ, χ become κ; π, β, φ become π; τ, δ, θ, ζ become σ. Examples:

φυλάττω, φύλακ-, φυλακτέος, ἀ-φύλακ-τος.
τάττω, ταγ-, εὐτάκτως 'in good order,' ἄ-τακ-τος 'disorderly.'
τειχίζω, τειχιδ- ἀτείχιστος 'unfortified.'
μέμφομαι, μεμφ- 'blame,' ἄ-μεμπ-τος 'blameless.'
So ἀν-υπόδη-τος 'unshod, barefoot,' δια-βα-τός 'fordable.'

-τέος GERUNDIVE

The verbal adjective in -τέος is used with the verb 'to be' expressing necessity, like the Gerundive in Latin. It can be used impersonally neuter, the verb taking its usual case; or with a noun as subject. The agent will be in the dative as in Latin. The neuter plur. is often used impersonally.

Impersonal. περὶ πολλοῦ ποιητέον (ἐστὶ) μηδεμίαν πρόφασιν τοῖς διαβάλλουσι δοθῆναι 'it is most important that no excuse should be given to slanderers'; ἀρκ-τέον ἡμῖν ἐστί 'we must rule.'

Personal. ὠφελητέα ἐστὶν ἡ πόλις 'the city must be helped.'

PARTICIPLES OF IMPERSONAL VERBS

These are used in the neuter impersonally instead of a clause or a genitive absolute. ἐξόν from ἔξεστι and παρόν from πάρεστι ('it being allowed'); δέον from δεῖ ('it being necessary'); ἀδύνατον ὄν 'since it is impossible'; δόξαν 'since it was decreed'; εἰρημένον 'since it has been stated,' etc.

ἐξὸν αὐτοῖς ἀπελθεῖν τί δῆτα μένουσι;
Why, pray, are they still here, when they might depart?

N.B. The Infinitive is used in certain adverbial expressions: ἑκὼν εἶναι 'willingly at least'; τὸ νῦν εἶναι 'at present'; ὡς εἰπεῖν 'so to speak'; (ὡς) ἐμοὶ δοκεῖν 'in my opinion.'

θεραπεύω serve, heal, court. ὁ ἱδρώς ἱδρωτ- sweat, toil.
ἀγαπάω love, am content to. τὸ ἄριστον breakfast.
ἡγέομαι lead (*gen.*); (*with clause*) think, judge.

Ex. 86. 1. τῶν γὰρ ὄντων ἀγαθῶν καὶ καλῶν οὐδὲν ἄνευ πόνου, ἐμοὶ δοκεῖν, καὶ ἐπιμελείας οἱ θεοὶ διδόασιν ἀνθρώποις· ἀλλ' εἴτε τοὺς θεοὺς ἵλεως εἶναί σοι βούλει, θεραπευτέον τοὺς θεούς· εἴτε ὑπὸ τῶν φίλων ἐθέλεις ἀγαπᾶσθαι, τοὺς φίλους εὐεργετητέον· εἴτε ὑπό τινος πόλεως ἐπιθυμεῖς τιμᾶσθαι, ἡ πόλις σοι ὠφελητέα ἐστί· εἴτε ὑπὸ πάσης τῆς Ἑλλάδος ἀξιοῖς θαυμάζεσθαι, τὴν Ἑλλάδα πειρατέον εὖ ποιεῖν· εἰ δὲ καὶ τῷ σώματι βούλει δυνατὸς εἶναι, γυμναστέον τὸ σῶμα σὺν πόνοις καὶ ἱδρῶτι.

2. ἆρα διδακτόν (a thing that can be taught) σοι φαίνεται ἡ ἀρετή;
3. οἱ δὲ πρέσβεις ἀπῆλθον ἄπρακτοι (without accomplishing anything).
4. εἷς δὲ λίθος ἐξαίρετος ἦν καὶ ὑπ' ἀνδρὸς ἑνός.
5. παρὸν γὰρ τῆς Ἀσίας ἄρχειν, ἄλλο τι αἱρήσεσθε;
6. τούτων δὲ γενομένων, ὡς ἐξὸν ἤδη αὐτοῖς ποιεῖν ὅ,τι βούλοιντο, πολλοὺς μὲν ἔχθρας ἕνεκα ἀπέκτεινον, πολλοὺς δὲ χρημάτων.
7. ἐπεμελεῖτο δὲ καὶ τούτου ὁ Κῦρος, ὅπως μὴ ἀν-ίδρωτοι γενόμενοι ἐπὶ τὸ ἄριστον καὶ τὸ δεῖπνον εἰσίοιεν. τοῦτο γὰρ ἡγεῖτο καὶ πρὸς τὸ ὑγιαίνειν (health) καὶ πρὸς τὸ ἡδέως ἐσθίειν ἀγαθὸν εἶναι, καὶ πρὸς τὸ δύνασθαι πονεῖν (endure).

Ex. 87. 1. Why are you so foolish as to stay here when you could (ἐξόν σοι) depart at once if you wish?
2. We must fortify the city since we dare not march out against so great an army.
3. "If I were allowed to choose," said I, "I would prefer to fight after breakfast."
4. Do not send the ambassadors away without-doing-any-business; we must try to make a treaty on those conditions.
5. Why is your city unfortified? Do you not consider that it would be (*p.* 78 (3)) much safer to wall it round?
6. For a long time I have been wanting (*pres. tense*) to go back to my country.
7. If we must retreat, let us march in good order.
8. If you had not left the gates unguarded, the city would not have been taken.
9. Since it is impossible to prevent him from departing, in my opinion we must try to find a fresh alliance.
10. He declared that, willingly at least, he would never injure one-who-had-helped the city so much.

ἵ-στη-μι 'I stand.' *Verb Stem* στα-. *Pres. Stem* ἱ-στα.

I. *Pres. Ind. Act.*	*Imperf. Act.*	*Fut. Act.*	*Pres. Mid. and Pass.*
ἵστημι	ἵστην	στήσω	ἵσταμαι
ἵστης	ἵστης	*Aor. Act.*	*Imperf.*
ἵστησι(ν)	ἵστη	ἔστησα	ἱστάμην
ἵσταμεν	ἵσταμεν	*Aor. Mid.*	*Perfect*
ἵστατε	ἵστατε	ἐστησάμην	ἕσταμαι
ἱστᾶσι(ν)	ἵστασαν		*Aor. Pass.*
Infin. ἱστάναι	*Subj.* ἱστῶ		ἐστάθην
Part. ἱστάς	*Opt.* ἱσταίην	*Imperat.* ἵστη, ἱστάτω etc.	

The Transitive Tenses of the Act. have the meaning 'set up.'

II. The Intransitive Tenses have the meaning 'stand.'

Perf. = Pres.	*Pluperf. = Imperf.*	*Imperative*
ἕστηκα	εἱστήκη	ἕσταθι, ἑστάτω, etc.
ἕστηκας	εἱστήκης	*Subjunctive*
ἕστηκε(ν)	εἱστήκει	ἑστῶ *or* ἑστήκω
ἕσταμεν	ἕσταμεν	*Optative*
ἕστατε	ἕστατε	ἑσταίην
ἑστᾶσι	ἕστασαν	*Infinitive* ἑστάναι
I am standing	I was standing	*Participle* ἑστηκώς *or* ἑστώς

ἑστώς, ἑστῶσα, ἑστός, Stem ἑστωτ-.

Aorist ἔστην 'I stood.' *Imperative* στῆθι, στήτω etc. *Subj* στῶ. *Opt.* σταίην. *Infin.* στῆναι. *Partic.* στάς (see p. 68).

Fut. στήσομαι; also mid. with an accusative.

COMPOUNDS

ἀνίστημι	set up	*Intrans. Tenses*	stand up.
ἀφίστημι	cause to revolt	„ „	revolt.
ἐφίστημι	put in charge	„ „	be in charge (*dat.*).
μεθίστημι	remove	„ „	change position.

OTHER PERFECTS WITH PRESENT MEANING

κέκτημαι, 'possess' from κτάομαι 'acquire.'
μέμνημαι, 'remember' from μιμνήσκω 'remind.'
εἴωθα 'am accustomed.' ἔοικα 'seem,' am 'like(ly).'
πέποιθα, 'trust' from πείθω 'persuade.'
πέφυκα, 'am by nature' from φύω 'produce.'
τέθνηκα 'am dead,' has shortened forms like ἕστηκα.
ἐγρήγορα, 'am awake' from ἐγείρω 'rouse.'
ἀπ-όλωλα, 'am ruined' from ἀπ-όλλυμι 'destroy.'

(ἡ) αὔριον morrow.
τὸ τρόπαιον trophy.
ἀργός -όν idle.
καθίστημι set up; *intrans. tenses* am appointed.
(ἡ) χθές yesterday.
ἀνθίσταμαι resist (*dat.*).
μισέω hate.

Ex. 88. 1. ἐπεὶ οἱ πολέμιοι ἀνεχώρησαν, τρόπαιον ἐστήσαμεν ὡς νενικηκότες.
2. ὁ δὲ προ-αισθόμενος τὰ αὐτὰ ταῦτα αὐτοὺς βουλευομένους, ἀποστῆναι ἀπὸ βασιλέως πρὸς τὸν Κῦρον, τοὺς μὲν ἀπέκτεινε τοὺς δὲ ἐξέβαλεν.
3. τί ἑστήκατε (ἕστατε) ἀργοὶ ἐν τῇ ἀγορᾷ πᾶσαν τὴν ἡμέραν;
4. ἐκέλευσε τοὺς παρ-εστῶτας συλλαβεῖν τὸν κλέπτην.
5. ἐπεὶ δὲ ἐτελεύτησε Δαρεῖος κατέστη εἰς τὴν βασιλείαν Ἀρταξέρξης.
6. τῇ δε αὔριον ἅμ' ἕῳ, ὥσπερ εἰώθεσαν, οἱ πολέμιοι ἀνέβησαν ἐπὶ τὸ ὄρος.
7. πολὺν χρόνον ἕστησαν σιγῶντες.
8. ὁ Θεὸς οὐδενὶ ἔοικεν· διὰ τοῦτο οὐδεὶς αὐτὸν ἐκμαθεῖν δύναται ἐξ εἰκόνος (εἰκών 'image').
9. ἐὰν δέ τις ἀνθιστῆται ἡμῖν πειρασόμεθα χειροῦσθαι τοῦτον (subdue).
10. ἅμ' ἡμέρᾳ ἀναστὰς ἐφύλαττε μὴ λάθοι αὐτὸν ὁ πατὴρ ἐξελθών.
11. πᾶς τις φυλάττειν ἐθέλει ἃ κέκτηται· οὕτω γὰρ πεφύκαμεν.
12. κατὰ τοὺς καθεστῶτας νόμους ἄξιοί ἐστε θανάτου.
13. ὀλώλαμεν, ὦ πολῖται, ὁ γὰρ στρατὸς νενίκηται, αἱ δὲ νῆες πόλυ ἀπέχουσιν.
14. μεταστήσωμεν τὴν πόλιν ἐξ ὀλιγαρχίας ἐς τὸ δημοκρατεῖσθαι.

Ex. 89. 1. "Halt, stranger," said the sentinel, "and say the watchword (σύνθημα), for otherwise it is not allowed to you to enter the camp."
2. Standing on the walls they threw down stones upon (κατά *with gen.*) us.
3. Those who were standing (οἱ *with perf. part.*) in the market-place, as it seems, were unable to escape.
4. They will all revolt from the alliance (συμμαχία) if we do not fight.
5. I have been appointed general in your place (ἀντί *with gen.*).
6. They are hated by all the Greeks on account of their not revolting from the rule of the Persians (*p.* 32).
7. If you had set this man over them they would have revolted.
8. Standing up he said that he hoped that all who were present would resist the unjust tyrant.
9. Those who possess much do not wish the laws to be changed.
10. I was standing in the market-place yesterday, as is my custom, trying to sell my goods; but no one would buy anything.
11. This custom (νόμος) was established (καθίστημι) in the days of Demosthenes, but in my opinion it ought to be changed.

THE DUAL

This number is sometimes used for twins and pairs of things, as 'two heads'; *δύο* however may always be used and the verb may be plural.

NOUNS.	1st *Decl.*	2nd *Decl.*	3rd *Decl.*	*ἐγώ*	*σύ*	*δύο*	*ἄμφω*
N.V.A.	-*α*	-*ω*	-*ε*	*νώ*	*σφώ*	*δύο*	*ἄμφω*
G. D.	-*αιν*	-*οιν*	-*οιν*	*νῷν*	*σφῷν*	*δυοῖν*	*ἀμφοῖν*

The article for all genders is *N.V.A.* *τώ*, *G. D.* *τοῖν* and the same applies to the remaining pronouns.

VERBS:	*Active and Aor. Pass.*		*Mid. and Pass.*	
	Primary	*Historic*	*Primary*	*Historic*
2.	-τον	-τον	-σθον	-σθον
3.	-τον	-την	-σθον	-σθην

Imperative Act. -τον -των *Mid. and Pass.* -σθον -σθων

Ex. 90. (*a*) καὶ νῦν δύο καλώ τε κἀγαθὼ ἄνδρε τέθνατον, καὶ οὔτε ἀνελέσθαι οὔτε θάψαι ἐδυνάμεθα.

(*b*) ἀδελφὼ ἐρίζοντε (quarrel) οὕτω διακεῖσθον ὥσπερ εἰ τὼ χεῖρε δια-κωλύοιεν ἀλλήλω.

(*c*) Φίλιππος, γενόμενος κριτὴς δυοῖν πονηροῖν, ἐκέλευσε τὸν μὲν φεύγειν ἐκ Μακεδονίας, τὸν δὲ ἕτερον διώκειν.

(*d*) *Socrates and two friends.*

καὶ ἐγὼ πρὸς τὸν Μενέξενον ἀποβλέψας (looking at), Ὦ παῖ Δημοφῶντος, ἔφην, πότερος ὑμῶν πρεσβύτερος; Ἀμφισβητοῦμεν (are in doubt), ἔφη. Οὐκοῦν (doubtless) ὁπότερος γενναιότερός ἐστιν, ἐρίζοιτε ἄν. Πάνυ γε, ἔφη. Καὶ μὴν ὁπότερός γε καλλίων, ὡσαύτως (likewise). ἐγελασάτην (γελάω I laugh) οὖν ἄμφω. Οὐ μὴν ὁπότερός γε, ἔφην, πλουσιώτερος ὑμῶν, οὐκ ἐρήσομαι (*vocab. p.* 72)· φίλω γάρ ἐστον. ἦ γάρ; (is it not so?) Πάνυ γε, ἐφάτην. Οὐκοῦν κοινὰ (common) τά γε φίλων λέγεται, ὥστε τούτῳ γε οὐδὲν διοίσετον (*fut. of* δια-φέρω differ), εἴπερ ἀληθῆ περὶ τῆς φιλίας λέγετον. συνεφάτην.

(*e*) ἴδεσθε τώδε τὼ κασιγνήτω (sisters), φίλοι,
ὣ τὸν πατρῷον οἶκον ἐξ-εσωσάτην·
τούτω φιλεῖν χρή, τώδε χρὴ πάντας σέβειν (reverence).

(*f*) ἀλλ' εἴπατόν μοι, σφὼ τίν' ἐστόν; — νώ; βροτώ.

VARIOUS RULES

(1) *οὐ μή* with the aor. subj. expresses a strong future denial: *οὐ μὴ πίθηταί σοι* 'he will certainly not obey you.'

(2) *οὐ μή* with a future interrogative expresses a strong prohibition: *οὐ μὴ σιγήσεις;* 'you must speak.'

(3) Verbs of hindering, denying, etc., often have *μή* with the infin.: and, if the verb itself is negatived, *μὴ οὐ* with the infin.: *εἴργει αὐτοὺς μὴ τοῦτο ποιεῖν* 'he prevents them from doing this,' but *οὐκ εἴργει αὐτοὺς μὴ οὐ τοῦτο ποιεῖν.*

(4) *πρίν* is used as a conjunction: (*a*) 'before,' (*b*) 'until.'

(*a*) If the main verb is affirmative *πρίν* ('before') has the infin.: *λέξαι θέλω σοι πρὶν θανεῖν ἃ βούλομαι* ('before I die').

(*b*) If the main verb is negative but in a Historic tense *πρίν* ('until') has the indic., both verbs stating a fact:

οὐ πρότερον ἐν-έδοσαν πρὶν οἱ ἱππῆς προσῆλθον.
They did not surrender until the cavalry arrived.

When *πρίν* is used indefinitely, *i.e.* referring to something which is in the future as regards the main verb, the construction is *ἄν* with aor. subj. after Primary, optative (without *ἄν*) after Historic tenses. See *μέχρι*, *ἕως*, etc. (p. 64 (2) and 66 (3)).

μὴ ἀπέλθητε πρὶν ἂν ἀκούσητε.
Do not depart until you have heard,

πρίν is also used as an adverb = formerly, before.

δημοσίᾳ publicly.	ἰδίᾳ privately.
ἐμποδών in the way.	ἐκποδών out of the way.
(ἐξ-)απατάω deceive.	ὁ ἕτερος the other (of two).

Ex. 91. 1. οὐ μή σε κρύψω πρὸς ὅντινα βούλομαι ἀφικέσθαι.
2. τί γὰρ ἐμποδὼν μὴ οὐκ ἀποθανεῖν αὐτούς;
3. νόμος ἐστὶν αὐτοῖς δημοσίᾳ μηδένα ἀποκτείνειν πρὶν ἂν εἰς Δῆλον ἀφίκηται τὸ πλοῖον.
4. πρὶν ἡμέραν εἶναι ἐς τὸν λιμένα εἰσέπλευσαν.
5. οὐ μὴ ἀπατήσῃς τὸν θεόν· οὐ μὴ πειράσει ἐξαπατᾶν αὐτόν;
6. φθάσας διῆλθε τὴν Θεσσαλίαν πρίν τινα κωλύειν.

Ex. 92. 1. Nothing hinders you from helping us privately.
2. You will certainly never defeat the king.
3. He died before he could finish the work.
4. Make (*mid.*) no one your friend until you know how he has treated (*χράομαι*) his former friends.
5. Let us not be angry until we have heard both speeches.
6. Publicly he offered to help us but privately he tried to prevent our departure.

OTHER VERBS IN -μι

τίθημι 'put, place,' follows *δίδωμι* very closely, substituting *τιθε-* for *διδο-* and *θε-* for *δο-*. Differences:

Imperf. Act. Sing. is *ἐτίθην, ἐτίθεις, ἐτίθει*. The subjunctives of *τίθημι* are normal. *Participles Pres. τιθείς Aor. θείς* are like **λυθείς**.

Pres. Act.	*Fut. Act.*	*Aor. Act.*	*Perf. Act.*	*Perf. Mid.*	*Aor. Pass.*
τίθημι	θήσω	ἔθηκα	τέθεικα	τέθειμαι	ἐτέθην
ἵημι	ἥσω	ἧκα	εἷκα	εἷμαι	εἵθην

Pres. Stem ἱε-, Verb Stem ἑ-, follows *τίθημι* closely.

Compounds of *τίθημι*: *προστίθημι* 'add'; *ἐπι-τίθε-μαι* 'attack.'

Compounds of *ἵημι*: *ἀφίημι* 'let go,' *συνίημι* 'understand.'

VERBS IN -*νυμι*, *e.g.* *δείκνυμι* 'point out'

Imperf. Act. *ἐδείκ-νυν, -νυς, -νυ.* *Imperat. 2nd Sing.* *δείκνῡ.* *Subj.* *δεικνύω.* *Opt.* *δεικνύοιμι.* *Partic.* *δεικ-νύς, -νῦσα, -νύν.* *Stem* *δεικνυντ-.*

In these verbs the tenses are formed from the stem, *e.g.* **δείκ-νυμι**, *δείξω, ἔδειξα* etc. like *φυλάττω* (p. 36).

Those with stem in -*ε* and -*ᾰ* have contracted future:

κρεμά-ννυμι	hang	*κρεμ-ά(σ)ω = κρεμῶ*	*ἐκρέμασα*
ἀμφιέννυμι	clothe	*ἀμφιέσω = ἀμφιῶ*	*ἠμφίεσα.*

Note also

open	break (*tr.*)	take oath
ἀν-οίγ-νυμι (ἀνοίγω)	ῥήγ-νυμι	ὄμ-νυμι
ἀνοίξω	ῥήξω	ὀμοῦμαι
ἀνέῳξα	ἔρρηξα	ὤμοσα
ἀνέῳχα	ἔρρωγα (*intr.*)	ὀμώμοκα
ἀνέῳγμαι, ἀνέῳγα	ἔρρηγμαι	
ἀνεῴχθην	ἐρράγην	ὠμόθην

The Pass. Pres. and Imperf. is formed like *δύναμαι* (p. 32) with **α ε ο υ** according to stem.

	Pres.	*Imperf.*
ἵημι	ἵεμαι	ἱέμην
δείκνυμι	δείκνυμαι	ἐδεικνύμην

but **κρεμάννυμι** has also **κρέμαμαι.**

θεῖναι νόμον to lay down a law, used of kings and tyrants.
θέσθαι νόμον to make a law, used of parliaments.

ἦ μήν assuredly.	ἁρπάζω seize.
ὁ νεκρός corpse.	δήπου doubtless.
ἀντ-έχω endure.	ὁ φονεύς murderer.

Ex. 93. 1. ὁ δ' Ἀλέξανδρος, ἁρπάσας κλίμακα παρὰ τοῦ φέροντος, προσέθηκε τῷ τείχει αὐτὸς καὶ ἀνέβαινεν.
2. ἀνοιχθείσης οὖν τῆς τοῦ δεσμωτηρίου θύρας ἀφῆκαν αὐτὸν ἐπεὶ ὤμοσε μὴ τοιαῦτα ποιήσειν ποτέ.
3. προσθεῖναι δὲ οὐδὲν εἶχον τοῖς εἰρημένοις οὐδ' ἀφελεῖν.
4. ὁ δὲ φονεύς, θεὶς τὸν νεκρὸν ἐν τῇ τραπέζῃ, πυρὶ ἧψε τὴν οἰκίαν ἐλπίσας δήπου, πάντων κατακαυθέντων, μηδένα ὑποπτεύσειν ὡς ἀποκτείνειε τὸν ἄνδρα.
5. τίθεται δέ γε τοὺς νόμους ἑκάστη ἡ ἀρχὴ πρὸς τὸ αὑτῇ συμφέρον, δημοκρατία μὲν δημοκρατικούς, τυραννὶς δὲ τυραννικούς, καὶ αἱ ἄλλαι οὕτω.
6. οἱ δὲ τοξόται κατέλαβον τὴν θύραν ἀνεῳγμένην.
7. χρόνον μὲν οὖν πολὺν ἀντεῖχον, οὐκ ἐνδιδόντες ἀλλήλοις· ἔπειτα δέ, ἦσαν γὰρ τοῖς Ἀθηναίοις οἱ ἱππῆς ὠφέλιμοι συμμαχόμενοι, τῶν ἑτέρων οὐκ ἐχόντων ἵππους, ἐτράποντο οἱ Κορίνθιοι καὶ ἀπεχώρησαν· καὶ ἔθεντο (piled) τὰ ὅπλα καὶ οὐκέτι κατέβαινον ἀλλ' ἡσύχαζον.
8. ἀφεὶς τὸν ἄνδρα ἄφες καὶ τὴν γυναῖκα αὐτοῦ.
9. ἀφιέμέν σε, ἐπὶ τούτῳ μέντοι, ἐφ' ᾧτέ σε ὀμεῖσθαι μηδέποτε ἀδικήσειν τοὺς φίλους.
10. ἐγὼ γάρ σε ταῦτα ἐδιδαξάμην (*p.* 30 (2)), ὅπως μὴ δι' ἄλλων ἑρμηνέων τὰς τῶν θεῶν συμβουλίας συνείης ἀλλ' αὐτὸς γιγνώσκοις.
11. ἡ τράπεζα κατερράγη.

Ex. 94. 1. Surely it is not safe to let a man go who has sworn to betray our city to the enemy. 2. He stopped (*aor. part. of* ἵστημι) his chariot in front of the troops (and) promised to give them more money. 3. Seizing the gold he opened the door and fled. 4. What would you do if you were to see a dead man hanging in the house? Personally, I should summon the police. 5. Do not put your feet on the table; whatever it may seem good to you to do in private, publicly such things are not done by gentlemen (καλὸς κἀγαθός). 6. We are pointing out the way to this man; but he, being a stranger, does not understand what we say. 7. If you add seven to five, what will it become? 8. When they opened the tomb, they found no corpse, for it had been opened before by thieves who took away everything. 9. Before he was hanged the murderer confessed that he had killed the judge for the sake of his money.

τὸ ἔαρ ἠρ- spring.
τὸ θέρος heat, summer.
τὸ φθινόπωρον autumn.
ὁ χειμών storm, winter.
ἡ ἄρκτος bear, north.
αἱ ἀνατολαί, ἡ ἕως, east.
ἡ μεσημβρία mid-day, south.
ἡ ἑσπέρα evening, west.
ὁ κύκλος circle.
ἐνθάδε here.

πάλιν back, εἰς τοὔμπαλιν (τὸ ἔμπαλιν) to the rear.

πάλαι long ago, οἱ πάλαι the ancients; πάλαι with present of verb = English perfect; with imperf. = pluperfect.

ἐν νῷ ἔχω, διανοέομαι, μέλλω = intend.

Ex. 95. 1. οἱ δὲ πολέμιοι ἐθαύμαζον ὅποι ποτὲ τρέψονται οἱ Ἕλληνες καὶ τί ἐν νῷ ἔχοιεν. ἐνταῦθα δὴ οἱ στρατηγοὶ πάλιν συνῆλθον καὶ συν-αγαγόντες τοὺς ἑαλωκότας ἠρώτων περὶ τῆς κύκλῳ πάσης χώρας τίς ἑκάστη εἴη. οἱ δ' ἔλεγον ὅτι τὰ μὲν πρὸς μεσημβρίαν τῆς ἐπὶ Βαβυλῶνα εἴη καὶ Μηδίαν, δι' ἧσπερ ἥκοιεν, ἡ δὲ πρὸς ἕω ἐπὶ Σοῦσά τε καὶ Ἐκβάτανα φέροι, ἔνθα θερίζειν καὶ ἐαρίζειν λέγεται ὁ βασιλεύς. ἡ δὲ διαβάντι τὸν ποταμὸν πρὸς ἑσπέραν ἐπὶ Λυδίαν καὶ Ἰωνίαν φέροι, ἡ δὲ διὰ τῶν ὀρῶν καὶ πρὸς ἄρκτον τετραμμένη ὅτι εἰς Καρδούχους ἄγοι.

2. Ξενοφῶν δὲ ἀπιὼν ἐπὶ τὸ εὐώνυμον ἀπὸ τοῦ δεξιοῦ ἔλεγε τοῖς στρατιωταῖς· Ἄνδρες, οὗτοί εἰσιν οὓς ὁρᾶτε μόνοι ἔτι ἡμῖν ἐμποδὼν τὸ μὴ ἤδη εἶναι ἔνθα πάλαι βουλόμεθα εἶναι.

3. Ἀντισθένης ἐπαινούμενός ποτε ὑπὸ πονηρῶν (rogues), Φοβοῦμαι ἔφη, μή τι κακὸν εἴργασμαι (ἐργάζομαι *deponent from* ἔργον 'deed').

Ex. 96. 1. The men of the present day are not worse than those of old.

2. Three times a month (*gen.*) they went to the temple of Zeus to sacrifice (θύω).

3. It was already evening and the king's army did not yet appear; and so many thought that he had gone back to his palace (βασίλεια *neut. pl.*).

4. We do not know what to-morrow will bring to us.

5. On these conditions therefore they gave their right (hands) (*fem.*) and took guarantees (πιστά *neut. pl.*).

6. In that country they give horns full of wine to their guests.

7. I advise that this man should be put out-of-the-way as quickly as possible that we may no longer have to guard against him.

8. In the winter I intend to go to the south where the sun always gives heat.

ζητέω seek, search for.
τὸ ψῦχος cold, coolness.
ἡ διάνοια mind.
ὅλος entire.
ἀρκεῖ it is sufficient.
βλέπω look, gaze.
φροντίζω think, ponder.
προσεύχομαι pray to.
τὸ ψήφισμα decree.
ἡ θεραπαινίς servant girl.
σκώπτω mock at.
ἐπι-δημέω live in a place.

Ex. 97. *The Eccentricities of Philosophers.*

1. Socrates as a soldier. συννοήσας (think of) γάρ ποτε πρόβλημά τι ἕω-θεν, ἐπειδὴ οὐ προὐχώρει (come out) αὐτῷ, εἱστήκει ζητῶν. καὶ ἤδη ἦν μεσημβρία, καὶ ἄνθρωποι ᾐσθάνοντο, καὶ θαυμάζοντες ἄλλος ἄλλῳ ἔλεγεν ὅτι Σωκράτης ἐξ ἑωθινοῦ (χρόνου) φροντίζων τι ἕστηκεν. τελευτῶντες δέ τινες τῶν Ἰώνων, ἐπειδὴ ἑσπέρα ἦν, δειπνήσαντες, καὶ γὰρ θέρος τότε γ' ἦν, χαμεύνια ἐξενεγκάμενοι, ἅμα μὲν ἐν τῷ ψύχει ἐκάθευδον, ἅμα δὲ ἐφύλαττον αὐτὸν εἰ καὶ τὴν νύκτα ἑστήξοι (*fut. perf. opt.*). ὁ δὲ εἱστήκει μέχρι ἕως ἐγένετο καὶ ἥλιος ἀνέσχεν. ἔπειτα ᾤχετο ἀπιὼν προσευξάμενος τῷ ἡλίῳ.

2. οὗτοι δέ που (generally speaking) ἐκ νέων πρῶτον μὲν εἰς ἀγορὰν οὐκ ἴσασι τὴν ὁδόν, οὐδὲ ὅπου δικαστήριον ἢ βουλευτήριόν ἐστι ἤ τι κοινὸν ἄλλο τῆς πόλεως συνέδριον (council)· νόμους δὲ καὶ ψηφίσματα, λεγόμενα ἢ γεγραμμένα, οὔτε ὁρῶσιν οὔτε ἀκούουσιν· καὶ τὰ εὖ καὶ τὰ κακῶς γεγονότα ἐν πόλει ὅλως αὐτοὺς λέληθεν (*p.* 80). καὶ ταῦτα πάντ' οὐδ' ὅτι οὐκ οἶδεν οἶδεν· ἀλλὰ τῷ ὄντι τὸ σῶμα μόνον ἐν τῇ πόλει κεῖται αὐτῶν καὶ ἐπιδημεῖ, ἡ δὲ διάνοια, ταῦτα πάντα ἡγησαμένη μικρὰ καὶ οὐδέν, πανταχῇ φέρεται, τά τε γῆς ὑπένερθε (beneath) καὶ τὰ ἐπὶ γῆς γεωμετροῦσα, οὐρανοῦ τε ὕπερ ἀστρονομοῦσα. ὥσπερ καὶ Θαλῆν, ἀστρονομοῦντα καὶ ἄνω βλέποντα, πεσόντα εἰς φρέαρ θεραπαινίς τις ἀπο-σκῶψαι λέγεται, ὡς τὰ μὲν ἐν οὐρανῷ μάλα βούλοιτο εἰδέναι, τὰ δὲ ἔμπροσθεν αὐτοῦ καὶ παρὰ πόδας λανθάνοι αὐτόν. ταὐτὸ δὲ ἀρκεῖ σκῶμμα ἐπὶ πάντας ὅσοι ἐν φιλοσοφίᾳ διάγουσιν.

3. οἱ δὲ φιλόσοφοι τοιοῦτοι σχολαστικοὶ ἐκαλοῦντο· εἷς δὲ τούτων ποτὲ οἰκίαν ἐθέλων πωλεῖν λίθον ἀπ' αὐτῆς εἰς δεῖγμα (sample) περιέφερεν.

A wish or regret for what is past or present is expressed by (1) the protasis of a conditional clause introduced by εἰ γάρ, εἴθε with Aor. Indic. for the past, Imperf. Indic. for the present time. Neg. μή.

(2) By ὤφελον -ες -ε etc., with Infinitive (sometimes preceded by ὡς).

E.g. εἰ γὰρ μὴ ἔπεμψα τὴν ἐπιστολήν 'Would that I had not sent the letter!'

ἀποδημέω to be away from home.
ἀπ-αντάω } ἐντυγχάνω } meet, happen upon (*dat.*).
ὁ πώγων πωγων- beard.
ἐπιλανθάνομαι forget (*gen.*), ἐπιλήσομαι, ἐπελαθόμην.
εἰκότως naturally, reasonably.
σχολαστικός -ή -όν learned, a pedant.

Ex. 98. 1. ἄλλῳ δέ ποτε σχολαστικῷ ἀποδημήσοντι φίλος εἶπεν· Ἀξιῶ (I beg) σε δύο δούλω ὠνεῖσθαί μοι, ἑκάτερον πεντεκαίδεκα ἐτῶν. ὁ δὲ ἀπεκρίνατο· Ἐὰν δὲ τοιούτους μὴ λάβω, ἀγοράσω σοι ἕνα τριάκοντα ἐτῶν.

2. σχολαστικός τις νεώτερος, ἐπεί τις εἶπεν, Ὁ πώγων σου ἤδη ἔρχεται, ἀπελθὼν εἰς τὴν πρὸς ἀνατολὰς πύλην (gate) ἐκεῖ εἱστήκει ἕως ἔλθοι ὁ πώγων. ἕτερος δὲ σχολαστικὸς ἐρωτήσας τί ποιεῖ καὶ γνούς, Εἰκότως, ἔφη, μῶροι καλούμεθα· πόθεν γὰρ οἶσθα εἰ διὰ τῆς ἑτέρας πύλης οὐκ ἔρχεται; (*N.B.* Neg. οὐ because this is not conditional but an indirect question).

3. σχολαστικῷ τινὶ φίλος ποτὲ ἀποδημῶν ἔγραψεν ἐπιστολὴν ἵνα βιβλία ἑαυτῷ ἀγοράσῃ. ὁ δέ, ἢ ἀμελήσας ἢ ὅλως ἐπιλαθόμενος τῶν προσταχθέντων, ἐπεὶ ἀπήντησε τῷ φίλῳ εἶπεν Τὴν ἐπιστολὴν ἣν ἔγραψας περὶ τῶν βιβλίων οὐκ ἐδεξάμην.

Note. ἐάω 'permit' has peculiarities: (1) Augment like ἔχω; *imperf.* εἴων, εἴας, εἴα. (2) Negative precedes it like φημί: οὐκ ἐῶ 'I forbid.' (3) Tenses formed by *Rule III*, *p.* 6: *fut.* ἐάσω, *aor.* εἴασα, *perf.* εἴακα.

Ex. 99. 1. We shall never permit such men to rule over us.

2. Would that I had never come to this place!

3. I am afraid that the city will be taken unless the allies come to the rescue quickly.

4. Do not be ashamed to confess that you forgot to post (ἀποπέμπειν) the letter.

5. If you had not let him go, he would not now be fighting against us.

6. For my part I do not think it right to allow children to do such things.

7. If ever you come to Athens I hope you will be my guest.

Don't forget your Aspirates!

ἰχνεύω track, hunt.
ἐκ-τρίβω wear out.
διάφορος different.
ἡ ἀλώπηξ fox.
ἐπί-κροτος -ον trodden hard.
ἡ ὁπλή hoof.

Ex. 100. ου το ιχνευειν την αλωπεκα εν τοις ορεσιν αλλα το αει επικρουειν επι τῃ επικροτῳ οδῳ, τουτ εστιν ο εκτριβει τας των ιππων οπλας.

It ain't the 'untin' on the 'ills that 'urts the 'orses' 'oofs but the 'ammer, 'ammer, 'ammer on the 'ard 'igh road. *Jorrocks.*

ἡ ἔκδοσις edition.
ἡ κλοπή burglary.
ἄγνωστος unknown.
τὸ κατάστημα store.
θραύω break open.
συνολικῆς altogether.
θρασύς bold.
πρωΐ, *adj.* πρωϊνός, early.
ἡ διασταύρωσις junction.
κ.κ. = κύριοι = Messrs.
ἡ ἐσχάρα fire-place.
τὸ εἶδος shape, instrument.

Ex. 101. *From a modern Greek newspaper*, ΕΜΠΡΟΣ.

ἐν Ἀθήναις. Σάββατον 8 ΟΚΤΩΒΡΙΟΥ. λεπτὰ 30.

ΤΡΙΤΗ ΕΚΔΟΣΙΣ.

ΘΡΑΣΥΤΑΤΗ ΚΛΟΠΗ ΕΙΣ ΤΗΝ ΟΔΟΝ ΣΤΑΔΙΟΥ.

τὰς πρωϊνὰς ὥρας τῆς χθές, ἄγνωστοι εἰσελθόντες εἰς τὸ ἐπὶ τῆς διασταυρώσεως τῶν ὁδῶν Σταδίου καὶ Ἐδουάρδου Λῶ κατάστημα ὀπτικῶν εἰδῶν τῶν κ.κ. Μαρινάρου, Μοιρανίου, καὶ Σιάσου, ἀφήρεσαν (ἀφεῖλον) πεντακισχιλίας δραχμάς, καὶ δισχίλια διάφορα ὀπτικὰ εἴδη, τηλεσκόπια, φωτογραφικὰς μηχανὰς κ.τ.λ. ἀξίας συνολικῆς πεντακισμυρίων δραχμῶν. οἱ κλέπται εἰσῆλθον εἰς τὸ κατάστημα ἐκ τοῦ ὑπο-γείου εἰς ὃ κατῆλθον θραύσαντες τὴν ἐσχάραν. ὡς ὕποπτοι συνελήφθησαν τρεῖς στρατιῶται.

20 εἴκοσι(ν).
30 τριάκοντα.
40 τετταράκοντα.
50 πεντήκοντα.
60 ἑξήκοντα.
70 ἑβδομήκοντα.
80 ὀγδοήκοντα.
90 ἐνενήκοντα.
100 ἑκατόν.
1000 χίλιοι.
10,000 μύριοι.
200 διακόσιοι.
300 τριακόσιοι.
400 τετρακόσιοι.
500 πεντακόσιοι.
600 ἑξακόσιοι.
700 ἑπτακόσιοι.
800 ὀκτακόσιοι.

900 ἐνακόσιοι. 30th τριακοστός -ή -όν.
30 times τριακοντάκις. 300th τριακοσιοστός.
300 times τριακοσιάκις; so χιλιοστός, χιλιάκις; μυριοστός, μυριάκις.

γαμέω 'I marry'; active with man as subject (Latin 'duco'), mid. with woman as subject, dative (Latin 'nubo').
Fut. γαμῶ, *Aor.* ἔγημα, *Perf.* γεγάμηκα.
Mid. γαμοῦμαι, ἐγημάμην, γεγάμημαι.

μνηστεύω	I woo.	ἐκδίδωμι	give in marriage.
καταλαμβάνω	catch up.	τὸ μῆλον	apple.

Ex. 102. *Atalanta's Wooing.*

There was once a king who had a very beautiful daughter. Many wooed her, but her father was unwilling to give her in marriage except (εἰ μή) to a man who could run very swiftly. For the girl herself surpassed all mortals in swiftness (ταχυτής, ταχυτῆτος *fem.*) of foot (*pl.*). Therefore he laid down such a law (*p.* 90) that, if she, pursuing, should be able to catch up the wooer (*pres. part.*), the man must die; but if not, she must marry him. Many therefore, either persuaded by love (ἔρως, ἐρωτ- *masc.*) of the girl or thinking that they could run faster, tried to win in this contest; but being defeated were put to death. But at last a certain Hippŏmĕnēs, helped by Aphrŏditē, was able to prevent her from catching him. For the goddess (θεά) gave him three golden apples; and, these being thrown (βάλλω, *p.* 54) behind him one by one (καθ' ἕν), while the girl each time (ἑκάστοτε) ceased running to gather (ἀναιρέω) them, the youth arrived first (φθάνω, *p.* 80). Thus therefore being victorious Hippomenes married the king's daughter. Afterwards they lived fortunately until the king died, and Hippomenes himself *succeeded to the throne* (Ex. 88, No. 5).

HINTS ON TRANSLATION

I. *Participles.* The use of these in English is restricted. We have really only two: *e.g.* from 'take,' Pres. Active, 'taking,' Past Passive, 'taken.' The ordinary Greek verb has eleven and some have more! How then are we to translate such a sentence as this?

ὁ δὲ στρατηγὸς ταῦτα ἀκούσας, πάντων ἤδη παρεσκευασμένων, πρῶτον μὲν ἐν ἀπορίᾳ ἦν· ἔπειτα δὲ λαβὼν τοὺς ἐκεῖ παρόντας ἐξεπορεύετο τιμωρησόμενος τοὺς ἐμβάλλοντας.

(1) *Translationese:* But the general having heard these things, all things having been already prepared, first was in perplexity; but afterwards, having taken those present there marched out about to punish those invading.

(2) *Translation:* Now when he had heard this, since everything was ready, the general was at first in doubt; afterwards taking all the available forces he set out to punish the raiders.

The first of two actions is expressed in Greek by an aorist participle even if the other verb is imperative; *e.g.* *λαβὼν ἴθι*, 'take it and go.' This aorist participle may often be rendered by the English present participle, especially if preceded by 'on,' 'after,' or 'by.'

II. *Neuter adjectives and pronouns.* Greek possesses a neuter gender so that an adj., partic. or pronoun can be put in the neuter and the reader left to supply a suitable noun. You must supply it. *ἠρώτησε πολλά* 'he asked many questions' (not 'he asked many things'). So *οὐδὲν ἀπεκρίνατο* 'he made no reply'; *δεινὰ ἔπαθεν* 'his sufferings were terrible'; *τοιαῦτα ποιῶν ὑπωπτεύετο* 'such conduct caused suspicion.'

III. Reversing a phrase or altering the order of clauses.

(1) A negative phrase will often be better translated by reversing.

οὐκ ἐδύνατο 'he failed.' *οὐ μόνον οὐ* 'so far from.'

(2) Passive for Active or vice versa.

ἦν δέ τις ἐν τῇ στρατιᾷ Ξενοφῶν, Ἀθηναῖος, ὃς οὔτε στρατηγὸς οὔτε λοχαγὸς ὢν συνηκολούθει, ἀλλὰ Πρόξενος αὐτὸν μετεπέμψατο οἴκοθεν 'but had been sent for by Proxenus from his home.' This is often necessary in order to get the antecedent close to the relative.

(3) To obtain emphasis in Greek, relative clauses are often placed early in the sentence and the antecedent follows later or is placed *in* the relative clause (*see* p. 20). For instance:

ὅστις δὲ εἴργασται ὥσπερ ἐγώ, πλέων καὶ κινδυνεύων, τί ἂν τις τοῦτον εἰς ἐκείνους τιθείη; εἰ μὴ τοῦτο λέγεις, ὡς ὃς ἄν σοι δανείσῃ, τοῦτον δημοσίᾳ μισεῖσθαι προσήκει

Why should one class with such as these a man who has earned money as I have by voyages and risks? Unless you mean to say that any one who lends you money deserves to be hated by the people.

IV. Many difficulties arise from the erroneous idea that the same part of speech must be used in English to represent each Greek word. Except in a very simple statement it is a good plan to practise translating into different parts of speech:

ὥσπερ εἰώθεσαν οἱ πάλαι πολεμοῦντες
According to the methods of ancient warfare.

πολλὴ ἀνάγκη ἐστὶ μεταλλάττειν τὰ πρὶν βεβουλευμένα
A change of policy is absolutely necessary.

V. Splitting up a long sentence.

οἱ δὲ ἐπειδὴ ἐν ταῖς ἀρχαῖς ἐγένοντο καὶ ἐξέτασιν ὅπλων ἐποιήσαντο, διαστήσαντες τοὺς λόχους ἐξελέξαντο τῶν τε ἐχθρῶν καὶ οἳ ἐδόκουν μάλιστα συμπρᾶξαι τὰ πρὸς τοὺς Ἀθηναίους ἄνδρας ὡς ἑκατόν, καὶ τούτων πέρι ἀναγκάσαντες τὸν δῆμον ψῆφον φανερὰν διενεγκεῖν, ὡς κατεγνώσθησαν, ἔκτειναν, καὶ εἰς ὀλιγαρχίαν τὰ μάλιστα κατέστησαν τὴν πόλιν

As soon as they were in office they held a review of the army. Stationing the companies some distance apart, they picked out their private enemies and any whom they believed to have co-operated with the Athenians. These came to about a hundred men. Immediately they compelled the people to condemn them publicly; and, as soon as they were condemned, put them to death. Thereupon they set up a government which was practically an oligarchy.

THE VARIOUS USES OF ὡς

(1) Introducing an indirect statement, like ὅτι:

ἔλεγεν ὡς αὐτοῖς μόνοις πατρὶς Πελοπόννησος *εἴη.*

(2) Purpose, like *ἵνα* and *ὅπως*, Subjunctive if main verb is primary, Optative if it is historic:

τούτου ἕνεκα ᾤετο δεῖσθαι φίλων ὡς συνεργοὺς ἔχοι

or with Fut. Participle, or before a preposition:

ἐξέπλευσαν ὡς ναυμαχήσοντες.
ἐξέπλευσαν ὡς ἐπὶ ναυμαχίαν.

(3) Past Time, with Indicative:

ὡς δὲ ἡ τροπὴ ἐγένετο, διασπείρονται.

(4) Cause, especially with a Participle; often a Gen. Absolute:

ἀγανακτοῦσιν ὡς μεγάλων τινῶν ἀπεστερημένοι.
ὡς πολέμου ὄντος παρ' ὑμῶν ἀπαγγελῶ;

(5) Exclamation: *ὡς ἀστεῖος ὁ ἄνθρωπος* ('how').
Wish: *ὡς ἀπόλοιτο καὶ ἄλλος* (not translated).

(6) Sometimes for *ὥστε* (consecutive), and *ὥσπερ* (comparative).

(7) As a preposition: towards, to a *person.*

ὡς τὸν Ἆγιν ἐπρεσβεύσαντο.

(8) As an adverb (*for οὕτως*) with accent:

ἀλλ' οὐδ' ὣς χαίροντες ἀπαλλάξαιτε ἄν.

(9) Making a parenthesis: *ὡς φασί* 'so they say';
ὡς εἰπεῖν 'so to speak.'

(10) Like 'ut' in Latin, without a verb:

ἦν δὲ οὐκ ἀδύνατος, ὡς Λακεδαιμόνιος, εἰπεῖν.

(11) With numerals ('about'): *παῖς ὡς ἑπταετής.*

(12) Strengthening adjectives and adverbs, especially superlatives:

ὑπερφυῶς ὡς χαίρω.
ὡς πλείστας ἤθροιζεν ἁμάξας.

WELL-KNOWN EXTRACTS FOR TRANSLATION AND REPETITION

Εἰς τοὺς ἐν Θερμοπύλαις θανόντας

ὦ ξεῖν', ἀγγέλλειν Λακεδαιμονίοις ὅτι τῇδε
κείμεθα, τοῖς 'κείνων ῥήμασι πειθόμενοι.

SIMONIDES.

Athens hears of Philip's approach

ἑσπέρα μὲν γὰρ ἦν, ἧκε δ' ἀγγέλλων τις ὡς τοὺς πρυτάνεις ὡς Ἐλάτεια κατείληπται. καὶ μετὰ ταῦθ' οἱ μὲν εὐθὺς ἐξαναστάντες μεταξὺ δειπνοῦντες, τούς τ' ἐκ τῶν σκηνῶν τῶν κατὰ τὴν ἀγορὰν ἐξεῖργον καὶ τὰ γέρρ' ἐνεπίμπρασαν, οἱ δὲ τοὺς στρατηγοὺς μετεπέμποντο καὶ τὸν σαλπιγκτὴν ἐκάλουν· καὶ θορύβου πλήρης ἦν ἡ πόλις. τῇ δὲ ὑστεραίᾳ, ἅμ' ἡμέρᾳ, οἱ μὲν πρυτάνεις τὴν βουλὴν ἐκάλουν εἰς τὸ βουλευτήριον, ὑμεῖς δ' εἰς τὴν ἐκκλησίαν ἐπορεύεσθε.

DEMOSTHENES.

The Sea—at last!

ἐπειδὴ δὲ οἱ πρῶτοι ἐγένοντο ἐπὶ τοῦ ὄρους καὶ κατεῖδον τὴν θάλατταν, κραυγὴ πολλὴ ἐγένετο. ἀκούσας δὲ ὁ Ξενοφῶν καὶ οἱ ὀπισθοφύλακες ᾠήθησαν ἔμπροσθεν ἄλλους ἐπιτίθεσθαι πολεμίους· ἐπειδὴ δὲ ἡ βοὴ πλείων τε ἐγίγνετο καὶ ἐγγύτερον καὶ οἱ ἀεὶ ἐπιόντες ἔθεον δρόμῳ ἐπὶ τοὺς ἀεὶ βοῶντας καὶ πολλῷ μείζων ἐγίγνετο ἡ βοὴ ὅσῳ δὴ πλείους ἐγίγνοντο, ἐδόκει δὴ μεῖζόν τι εἶναι τῷ Ξενοφῶντι. καὶ ἀναβὰς ἐφ' ἵππον καὶ τοὺς ἱππέας ἀναλαβὼν παρεβοήθει· καὶ τάχα δὴ ἀκούουσι βοώντων τῶν στρατιωτῶν, Θάλαττα, θάλαττα, καὶ παρεγγυώντων. ἐπεὶ δὲ ἀφίκοντο πάντες ἐπὶ τὸ ἄκρον, ἐνταῦθα δὴ περιέβαλλον ἀλλήλους καὶ στρατηγοὺς καὶ λοχαγοὺς δακρύοντες. οἱ δὲ στρατιῶται φέρουσι λίθους καὶ ποιοῦσι κολωνὸν μέγαν.

XENOPHON.

The miser and the mouse

μῦν Ἀσκληπιάδης ὁ φιλάργυρος εἶδεν ἐν οἴκῳ,
καί, τί ποιεῖς, φησίν, φίλτατε μῦ, παρ' ἐμοί;
ἡδὺ δ' ὁ μῦς γελάσας, μηδέν, φίλε, φησί, φοβηθῇς·
οὐχὶ τροφῆς παρὰ σοὶ χρῄζομεν, ἀλλὰ μονῆς.

LUCILIUS.

Injustice

SOCRATES. *μέγιστον τῶν κακῶν τυγχάνει ὂν τὸ ἀδικεῖν.*

POLUS. *ἦ γὰρ τοῦτο μέγιστον; οὐ τὸ ἀδικεῖσθαι μεῖζον;*

SOC. *ἥκιστά γε.*

POL. *σὺ ἄρα βούλοιο ἂν ἀδικεῖσθαι μᾶλλον ἢ ἀδικεῖν;*

SOC. *βουλοίμην μὲν ἂν ἔγωγε οὐδέτερα· εἰ δ' ἀναγκαῖον εἴη ἀδικεῖν ἢ ἀδικεῖσθαι, ἑλοίμην ἂν μᾶλλον ἀδικεῖσθαι ἢ ἀδικεῖν.*

POL. *σὺ ἄρα τυραννεῖν οὐκ ἂν δέξαιο;*

SOC. *οὔκ, εἰ τὸ τυραννεῖν γε λέγεις ὅπερ ἐγώ.*

POL. *ἀλλ' ἔγωγε τοῦτο λέγω ὅπερ ἄρτι, ἐξεῖναι ἐν τῇ πόλει, ὃ ἂν δοκῇ αὐτῷ, ποιεῖν τοῦτο, καὶ ἀποκτείνοντι καὶ ἐκβάλλοντι καὶ πάντα πράττοντι κατὰ τὴν αὑτοῦ δόξαν.*

PLATO.

Retreat

καύσαντες οὖν πυρὰ πολλὰ ἐχώρουν οἱ Ἀθηναῖοι ἐν τῇ νυκτί. καὶ αὐτοῖς, οἷον φιλεῖ γίγνεσθαι καὶ πᾶσι στρατοπέδοις, μάλιστα δὲ τοῖς μεγίστοις, φόβοι καὶ δείματα ἐγγίγνεσθαι, ἄλλως τε καὶ ἐν νυκτί τε καὶ διὰ πολεμίας καὶ ἀπὸ πολεμίων οὐ πολὺ ἀπεχόντων ἰοῦσιν, ἐμπίπτει ταραχή· καὶ τὸ μὲν Νικίου στράτευμα, ὥσπερ ἡγεῖτο, συνέμενέ τε καὶ προὔλαβε πολλῷ, τὸ δὲ Δημοσθένους, τὸ ἥμισυ μάλιστα καὶ πλέον, ἀπεσπάσθη τε καὶ ἀτακτότερον ἐχώρει. ἅμα δὲ τῇ ἕῳ ἀφικνοῦνται ὅμως πρὸς τὴν θάλατταν.

THUCYDIDES.

Ὄρνις χρυσοτόκος

ὄρνιθος ἀγαθῆς ὠὰ χρυσέα τικτούσης,
θησαυρὸν ᾠέθ' ὁ δεσπότης ἀνευρήσειν
κἄκτεινε ταύτην, ἀθρόον λαβεῖν μέλλων.
εὑρὼν δ' ὅμοια τἄνδον ὀρνέοις ἄλλοις,
ᾤμωζε πλεῖστον ἐλπίδων ἀτευκτήσας·
πλείονος ἔρως γὰρ ἐστέρησε τῶν ὄντων.

BABRIUS.

TABLE OF CONDITIONAL SENTENCES

The only difficulty of these lies in English use of tenses.

I. When the speaker assumes that if a certain act is done in the past, present, or future, the result (has) followed, is following or will follow automatically:

(*a*) If they have crossed (= are across), they have escaped.
si transierunt, effugerunt.
εἰ διαβεβήκασιν, ἐκπεφεύγασιν.

(*b*) If they are crossing the river, they are escaping.
si flumen transeunt, effugiunt.
εἰ ποταμὸν διαβαίνουσιν, ἐκφεύγουσιν.

(*c*) If they (shall) cross, they will escape.
si transibunt, effugient.
εἰ διαβήσονται, ἐκφεύξονται.

Any combination of tenses is possible and the main clause (apodosis) may be a command, or a wish: *e.g.*

If they are crossing the river, let them be recalled.

II. When the speaker discusses the possibility of the event and its result, *i.e.* in some future time:

(*a*) If they cross, they will escape.
si transierint (*fut. perf.*), effugient.
ἐὰν διαβῶσι, ἐκφεύξονται.

Here the speaker is certain about the result only; in I (*c*) he is certain both of the act and of the result (and here (*b*) he is vague about both).

(*b*) If they {should cross / were to cross / crossed}, they would escape.
si transeant, effugiant.
εἰ διαβαῖεν, ἐκφύγοιεν ἄν.

III. When the speaker considers what the result might have been, if things had happened or were happening differently.

(*a*) If they had crossed, they would have escaped.
si transiissent, effugissent.
εἰ διέβησαν, ἐξέφυγον ἄν.

(*b*) If they {were / should be} now crossing, they would (now) be escaping.
si (nunc) transirent, effugerent.
εἰ διέβαινον, ἐξέφευγον ἄν.

III (*a*) and (*b*) may be combined, *e.g.*

If they had crossed, they would now be escaping.
si transiissent, nunc effugerent.
εἰ διέβησαν, *(νῦν) ἐξέφευγον ἄν.*

Notice then that the English 'if they cross' may represent either a subjunctive or a future; and that 'if they crossed' may represent an imaginary case in the future!

SENTENCES FOR REVISION

The references to rules may be used while the sentences are being done or afterwards for correction. The figures in parentheses refer to pages.

PART I

1. We told him not to trust (5) that man but the words of the young man persuaded him to send the gold away (15).

2. The ships which we have in the river will be (12, *Rule VI*) much (18 (4)) more useful to you than to us (18 (2)) since our sailors are not present.

3. This general however is said to be so clever that (24) he has won many battles both by land and sea.

4. Some were ordered (27) to prepare the ships for a sea-fight; others to destroy the bridge (6 (3)) which the enemy had made (23, *Note*).

5. "We have not so many soldiers as the enemy have (22)," said he, "but we can trust those men that we have with us (20 (2))."

6. Instead of wishing (32) to be a rich man you ought to consider it of great importance (31) to have wisdom and virtue.

7. He said that he himself (28) was too wise to trust (28) these men but that many others had been injured by them.

8. The burglars were carrying away (30) much gold and beautiful gifts because the servants had not shut the doors of the poet's house.

9. For a long time we remained quiet (4, *Rule II*) in the village but on the third day (16) we marched through the plain to the river.

10. Ought we not to arm the citizens (16) since those barbarians who have broken the treaty are not far distant from this place (16)?

11. On the same day two great battles, one on land, and the other (11, *Rule V*) on the sea, checked the Persians who were attacking our land.

PART II

1. As soon as (64) they had (23, *Note*) found the waggons which they had left behind, they consulted together, being doubtful what they ought to do next.

2. My son, fear God; honour the king; love your father and your mother.

3. You have been chosen judge not that (62) you may please yourself but that wrong-doers (44 (5)) may be justly punished.

4. Some prepared the ships for battle, others were sent to build again (62) the walls which were damaged by the enemy.

5. For how much will you sell me this house? Do not demand (60) a very great price for I have very little money.

6. When he was rich (64) many wished to become his friends; but now that he is poor very few are willing to help him.

7. When a battle takes place (64 (2)), those who are fighting for their country fight better than those who are invading the land for gold.

8. The Persian cavalry always fought in this manner: whenever the enemy attacked (66 (3)) they themselves used to retreat; but when the enemy in their turn retreated, they attacked them from all sides.

9. Why did you try to conceal from us the source (36) of the money which you received? This concerns me as well as you (56).

10. Let the men who did this be punished (60); and may all who do the like (66 (1)) be put to death! and the sooner the better (Ex. 64, No. 4).

11. Throwing away their weapons they promised to follow the general wherever he might lead them (66 (3)).

PART III

1. If 20 men can build a wall in 30 days, in how many days can 100 men make the same wall? Come now tell me this (60).

2. What would have happened if Alexander (76 (*b*)), after subduing the Persian empire, had crossed into Italy to fight the Romans? In my opinion, he would have overcome them owing to his having better cavalry (32).

3. By starting at nightfall they crossed the river without being noticed by the enemy's sentinels and reached the city first (80).

4. As soon as he perceived that some were crossing the bridge (68) but that many were left on the other side he fell upon those-who-had-crossed with all the forces he could muster before they could form a line (89 (4)).

5. Since he never uses the books he has already, I should not be acting wisely if I sent him more (78).

6. I asked you these questions in order to know whether you remember what you read in this book.

7. I am afraid that he has deceived us (74); for I happened to be present (80) at the time when (64 (1)) he was trying to persuade the people to revolt. Still, let us not be despondent.

8. Do not cease to fortify the city (60) until you have heard (89) whether it is expedient for us (56) to surrender or not.

9. Our soldiers were plainly (80) too weak to resist their attack since (44 (1)) they had marched all night during a great storm (42 (1)).

10. Of all the villages which (20) I saw on my travels I consider this one to be the most beautiful in spite of its small size (44 (2)).

11. Insert suitable *particles* at the points marked.

The * king spoke thus to his troops. Our * general, more wisely *, said a few words only as follows: Before * you is the enemy; behind * flows an impassable river. You * must conquer or die. I do * not fear that you will surrender without a great struggle; I * know that you have often defeated greater armies than these.

GREEK-ENGLISH INDEX

ἀγαθός good, brave
ἀγαπάω I love
ἀγγέλλω I announce
ἄγγελος messenger
ἁγιάζω I keep holy
ἀγορά market-place
ἀγοράζω I buy
ἀγορεύω I make a speech
ἄγω I lead
ἀγών contest, game
ἀδελφός brother
ἀδικέω I injure
ἄδικος unjust
ἀδύνατος impossible, unable
ἀεί always
ἀθάνατος immortal
Ἀθῆναι Athens
Ἀθηναῖος Athenian
ἀθλητής athlete
ἀθρόος whole, all at once
ἄθυμος despondent
Αἴγυπτος Egypt
αἴνιγμα riddle
αἴρω I lift, set out
αἰσχρός ugly, disgraceful
αἰσχύνομαι I am ashamed
αἰτέω I beg
ἀκούω I hear
ἄκρος top, edge
ἄκων unwilling(ly)
ἀλ- 66
ἀλγεινός painful
Ἀλέξανδρος Alexander
ἀλήθεια truth
ἀληθής true
ἀλλά but
ἀλλήλους each other
ἄλλο else
ἄλλος other, else
ἀλλότε at another time
ἄλλως otherwise
ἄλλως τε καί especially
ἅμα together
ἅμαξα wagon
ἀμαχεί without a fight
ἀμείνων better
ἀμελέω I neglect
ἀμφί about
ἄμφω both
ἀνά up
ἀναγιγνώσκω I read
ἀνατείνω stretch out
ἀνατολή the east
ἀναχωρέω I retreat
ἀναχώρησις a retreat
ἀνδρεῖος brave
ἀνεξεύρετος incalculable
ἄνευ *prep.* without
ἀνέχω I hold out
ἀνήρ (ἀνδρ-) man
ἀνθ᾽ *see* ἀντί
ἄνθρωπος man
ἀνίδρωτος without toil
ἀνίστημι I raise up
ἀνοίγνυμι I open
ἄνομος lawless
ἀντί instead of
ἄνω up
ἀξιόπιστος believable
ἄξιος worth, worthy
ἀξιόω I claim, think worthy
ἀπαιτέω I demand
ἅπας all, whole
ἀπελθ- depart
ἀπό from
ἀποδημέω I am away
ἀποδίδομαι I sell
ἀποδιδράσκω I run away
ἀποθνήσκω I die, am killed
ἀποκτείνω I kill
ἀπόλλυμι I destroy
ἀπορέω I am perplexed
ἀποσπάω I separate
ἀποστέλλω I send off
ἀποστερίσκω I deprive
ἄρα truly, therefore
ἆρα *question-word*
ἀργός idle
ἄργυρος silver
ἀρετή virtue, courage
ἀριθμός number
ἄριστον breakfast
ἄριστος best
Ἀρκάς an Arcadian
ἅρμα chariot
ἄρτι just now, lately
ἄρτος bread
ἀρχή rule, beginning
ἄρχομαι I begin
ἄρχω I rule
ἀστήρ star
ἀσφαλής safe
ἄτακτος disorderly
ἅτε since, because

ἀτείχιστος unfortified
ἀτευκτέω I miss, lose
αὖ, αὖθις again
αὔριον to-morrow
αὐτίκα immediately
αὐτός self 41
ἀφ' *see* ἀπό
ἀφαιρέω I take away
ἀφίημι I let go, forgive
ἀφικνέομαι I arrive
ἀφίστημι I cause to revolt

βαδίζω I walk, march
βάκτρος staff
βάλλω I throw, pelt
βάρβαρος foreign
βασιλεία kingdom
βασιλεύς king
βασιλικός kingly
βάσις base
βαΰζω I bark at
βελτίων better
βιβλίον book
βίβλος roll, book
βίος life
βλάπτω I hurt
βλέπω I look
βοάω I shout
βοή a shout
βοηθέω I go to help
βουλευτήριον council-room
βουλεύω consult
βουλή council
βούλομαι I desire
βοῦς ox, cow
βροτός mortal

γαμέω I marry
γάρ *conj.* for
γαστήρ belly
γε at any rate
γεν- *see* γίγνομαι
γενναῖος adj. noble
γέρρον hut
γέρων old man
γεύομαι I taste
γεωμετρέω I measure
γεωργός farmer
γῆ earth, land
γίγνομαι I become, happen
γιγνώσκω I get to know
γραῦς old woman
γραφεῖον brush, pencil
γράφω I write
γυμνάζω I train
γυμνάσιον gymnasium
γυνή woman, wife

δάκνω I bite
δακρύω I weep
δαρεικός gold coin
Δαρεῖος Darius
δέ but, and
δέδοικα I fear
δεῖ it is necessary
δείκνυμι point out
δειλός cowardly
δεῖμα terror
δεινός terrible, clever at
δεῖπνον dinner
δέκα ten
Δελφοί oracle of Apollo
δένδρον tree
δεξιός right-hand
δέομαι I beg, need
δέος fear
δεσμός chain
δεσμωτήριον prison
δεσμώτης prisoner
δεῦρο / δεῦτε come hither
δεύτερος second
δέχομαι I receive
δή 36
δηλόω I make plain
δημαγωγός popular leader
δῆμος people
δήπου doubtless
δῆτα 36
Δία 82
διά *acc.* on account of, *gen.* through
διαβαίνω I cross
διάγω I pass the time
διακεῖμαι to be in a state
διανοέομαι I intend
διάνοια mind
διατελέω I continue
διαφθείρω I destroy
διάφορος different
διδάσκω I teach
δίδωμι I give
δίκαιος just
δικαστήριον law-court
δίκη law, penalty
διό wherefore
διότι because
διφθέρα skin coat
διψάω I thirst
διώκω I pursue
δο- 76
δοκεῖ it seems (good)
δοκέω I seem (think)
δόξα opinion, fame
δουλεία slavery

δοῦλος slave
δουλόω I enslave
δραχμή drachma (10*d*.)
δράω I do, act
δρᾶμα action
δρόμος running
δύναμαι I am able
δύναμις power
δύο two
δώδεκα twelve
δῶρον gift

ἔαρ spring (season)
ἑαυτόν himself
ἐάω I allow
ἕβδομος seventh
ἐγγύς near
ἐγώ I
ἐθέλω I am willing
εἰ if, whether
εἶ thou art, thou goest
εἶδον I saw
εἴθε would that...!
εἴκοσι twenty
εἰμί I am
εἶμι I am going
εἶναι to be
εἴπερ if indeed
εἶπον I said
εἴργω I prevent
εἴρηκα I have said
εἰς (ἐς) into, for
εἷς one
εἴτε whether...or
ἐκ out of, from
ἕκαστος each
ἑκάτερος each of two
ἐκεῖ there
ἐκεῖθεν thence
ἐκεῖνος 20
ἐκεῖσε thither
ἐκκλησία assembly
ἐκποδών out of the way
ἕκτος sixth
ἑκών willing(ly)
ἑλ- 66
ἐλάττων less
ἐλαύνω I drive, march
ἐλέφας elephant, ivory
ἐλθ- come, go, 70
Ἑλλάς Greece
Ἕλλην a Greek
ἐλπίζω I hope, expect
ἐλπίς hope
ἐμός my, mine
ἐμπίπρημι I set on fire
ἐμποδών *adv.* in the way
ἔμπροσθεν in front of
ἐμφαίνω I emphasise
ἐν in, among, on
ἕνδεκα eleven
ἕνεκα for the sake of
ἔνθα where, whereupon
ἐνθάδε here
ἐνιαυτός year
ἔνιοι some
ἐνίοτε sometimes
ἐννέα nine
ἐνοικέω I inhabit
ἐνταῦθα then, there
ἐξ = ἐκ
ἕξ six
ἐξαίρετος removable
ἐξενεγκ- carry out
ἔξεστι it is allowed
ἐξ οὗ since
ἐπαινέω I praise
ἐπεί when, since
ἐπειδή when, since
ἔπειτα afterwards
ἐπί + *acc.* against
ἐπιδημέω I stay at home
ἐπιθυμέω I desire
ἐπιλαθ- forget
ἐπιλείπω fail
ἐπιμέλεια care, pains
ἐπιμελέομαι I take care of
ἐπιούσιος daily
ἐπίσταμαι I know how to
ἐπιστολή letter
ἐπιτίθεμαι I attack
ἕπομαι I follow
ἑπτά seven
ἔργον work, deed
ἔρημος uninhabited
ἐρίζω I quarrel
ἔρις strife
ἑρμηνεύς interpreter
ἔρχομαι I go
ἐρῶ I shall say
ἔρως love
ἐρωτάω I inquire
ἐς = εἰς
ἐσθίω I eat
ἑσπερά evening, west
ἔστε until
ἕστηκα I stand
ἔσχατος last
ἕτερος the other
ἔτι still, yet
ἔτος year
εὖ *adv.* well

εὐγενής well-born
εὐεργετέω I benefit
εὐθύς immediately
εὐκλεής famous
εὔλογος reasonable
εὑρίσκω I find
εὖρος breadth
εὐτυχής fortunate
εὐώνυμος left-hand
ἐφ' = ἐπί
ἔφασαν they said
ἔφη he said
ἐφίστημι I set over
ἐφ' ᾧτε on condition that
ἐχθρός hostile
ἔχω I have 22
ἕωθεν at dawn
ἕως dawn
ἕως while, until

ζάω I am alive
ζημία penalty
ζημιόω I fine, punish
ζητέω I seek
ζώνη belt
ζῷον animal

ἦ *adv.* truly
ἤ either, or, than
ᾗ by what road, where
ἡγέομαι I lead, think
ἥδε this (*fem.*)
ἡδέως gladly
ἤδη already, soon
ἥδομαι I am pleased
ἡδονή pleasure
ἡδύς sweet
ἤθελον I wished
ἥκιστα certainly not
ἥκω I am come
ἦλθον I went, came
ἥλιος sun
ἡμεῖς we
ἡμέρα day
ἡμέτερος our
ἥμισυ half
ἤν if
ἤρετο he asked
Ἡρόδοτος Herodotus
ἦρος in spring
ἥρως hero, demigod
ἡσυχάζω I keep quiet
ἥσυχος quiet
ἥττων less
ἠχώ echo

θάλαττα sea
θάνατος death
θάπτω I bury
θαυμάζω I wonder
θεά goddess
θέλημα will
θέλω I wish
θεός god
θεραπεύω serve, heal
Θερμοπύλαι Thermopylae
θέρος heat, summer
θέω I run
Θῆβαι Thebes
θήρ, θηρίον wild beast
θηράω, θηρεύω } I hunt
θησαυρός treasure
θνητός mortal
θόρυβος uproar
Θρᾷξ Thracian
θυγατήρ daughter
θύρα door
θύω I sacrifice

ἰατρός physician
ἰδ- see
ἱδρώς sweat
Ἰησοῦς Jesus
ἵλεως propitious
ἵνα 62
Ἰουδαῖος Jew
ἱππεύς horseman
ἱππεύω I ride
ἱπποπόταμος hippopotamus
ἵππος horse
ἴσος equal
ἰσχυρός strong
ἴσως perhaps
ἰχθύς fish

καθεστώς established
καθεύδω I sleep
κάθημαι I sit
καθίστημι I appoint
καινός new, fresh
καίπερ although
καίτοι and yet
καίω I burn
κακός bad, cowardly
κακόω I harm
κακῶς ἔχειν to be unfortunate
καλέω I call
καλός beautiful, noble
καλῶς nobly, well
κανών rod

κατά along, according to
καταγιγνώσκω I condemn
κατακεῖμαι I lie down
κατακρίνω I condemn
καταλαμβάνω I capture
καταπίπτω I fall off
καταφρονέω I despise
κατοικέω I inhabit
κάτω down
κεῖμαι I lie on
κελεύω I order
κέρας horn, wing of army
κεφαλή head
κῆρυξ herald
κηρύττω I proclaim
κιβώτιον box
κινέω I move
κλείω I shut
κλέπτης thief
κλέπτω I steal
κλῖμαξ ladder
κοινός *adj.* common
κολάζω I punish
κολωνός cairn
κομίζω I convey
κόπτω I cut, knock
κόρη girl
Κόρινθος Corinth
κραυγή shout
κρέμασθαι to hang
κρίνω I judge
κριτής judge
κροκόδειλος crocodile
κρούω strike, knock
κρύπτω I conceal
κτάομαι acquire
Κῦρος Cyrus
κύων dog
κωλύω I prevent
κώμη village

λαγχάνω I obtain
Λακεδαιμόνιος Spartan
λαμβάνω I take
λανθάνω 80
λέγω I say
λείπω I leave
λεπτόν small coin
λέων lion
λήθη forgetfulness
λιμήν harbour
λιπ- leave
λιτανεία prayer
λόγος word, speech
λοιπόν the future
λοιπός remaining
λούω I wash
λοχαγός captain
λύγξ lynx
Λυκοῦργος Lycurgus
λύομαι I ransom
λυρικός *adj.* lyric
λυσιτελεῖ it is profitable
λύω I loose

μαθηματική mathematics
Μακεδών a Macedonian
μάλα much
μάλιστα especially
μᾶλλον more, rather
μανθάνω I learn
μάχη battle
μάχομαι I fight
με me (*acc.*)
μέγας great
μεθ' = μετά
μείζων greater
μέλας black
μέλει it concerns
μέλι honey
μέμνημαι I remember
μέμφομαι I blame
μέν on the one hand
μέντοι however
μένω I remain, await
μεσημβρία mid-day, south
μέσος middle
μετά *acc.* after, *gen.* with
μεταγιγνώσκω I repent
μεταξύ *adv. and prep.* between
μεταπέμπομαι I send for
μέχρι (οὗ) until
μηδείς nobody
μήν month
μήν *adv.* truly
μηχανάομαι I contrive
μικρός small
μισέω I hate
μισθός pay, reward
μνᾶ mina £4
μοι me (*dat.*)
μονή lodging
μόνον *adv.* only
μόνος alone, only
μουσική music
μῦς mouse
μῶρος foolish

ναυαγέω I suffer shipwreck
ναυαρχέω I command a ship

ναυμαχία sea-fight
ναῦς ship
ναύτης sailor
νεανίας young man
Νεῖλος the Nile
νεκρός corpse
νέκταρ nectar
νεός new, young
νέω I swim
νῆσος island
νικάω I conquer
νίκη victory
νομίζω I think, consider
νόμος custom, law
νόος (νοῦς) mind, brains
νόσος disease
νῦν now
νύξ night

ξεῖνος } guest, stranger
ξένος }
ξύλον timber

ὄγδοος eighth
ὅδε this
ὁδίτης traveller
ὁδός road, way
οἶδα I know, 68
οἴκαδε homewards
οἰκεῖος domestic
οἰκία house
οἰκοδομέω I build
οἴκοι at home
οἰκονομία management
οἶκος house
οἰκτείρω I pity
οἰμώζω I bewail
οἶνος wine
οἴομαι (οἶμαι) I think
οἷος such as
οἴχομαι I am gone
ὀκτάπους } octopus
ὀκτώπους }
ὀλιγαρχία rule of few
ὀλίγος little, *pl.* few
ὅλος entire
Ὅμηρος Homer
ὄμνυμι I take oath
ὅμοιος like
ὁμοιότης likeness
ὁμολογέω I confess
ὅμως still, nevertheless
ὄνομα name
ὄντι, τῷ really
ὄνυξ hoof

ὅπῃ where
ὄπισθεν behind
ὁπλίζω I arm
ὅπλον weapon
ὅποι whither
ὁπότε whenever
ὁπότερος which (of two)
ὀπτικός optical
ὅπως 62
ὁράω I see
ὀργίζομαι I am angry
ὁρμάω I rush
ὄρνις bird, hen
ὄρος mountain
ὀρχήστρα dancing-floor
ὅς *rel.* who, which
ὅσος (as great) as
ὅσπερ (the one) who
ὀστέον bone
ὅστις whoever, who
ὅτε 64
ὅτι that, because
ὅτου, ὅτῳ, *see* ὅστις
οὐ, οὐκ, οὐχ, οὐχί not
οὐδαμ- 36
οὐδέ not even
οὐδείς nobody
οὐδέν nothing, in no way
οὐδέποτε never
οὐδέπω not yet
οὐδέτερος neither
οὐκέτι no longer, no more
οὖν therefore
οὐρά tail
οὐρανός sky, heavens
οὔτε neither, nor
οὗτος this, that
οὕτως, οὕτω thus, so
ὀφειλέτης debtor
ὀφείλημα debt
ὄψομαι I shall see

παθ- suffer
παιδεύω I train, educate
παῖς boy, girl, son
πάλαι long ago
πάλιν *adv.* back
πάνθηρ leopard
πανταχόσε in all directions
πανταχοῦ everywhere
πάνυ quite, certainly
παρά 12
παρασάγγης = 30 stadia
παρεγγυάω I pass on orders
πάρειμι I am present

πάρειμι I pass by
πᾶς all, whole, every
πάσχω suffer, experience
πατήρ father
πατρίς fatherland
παύομαι I cease
παύω I check
πεδίον a plain
πείθομαι I obey
πείθω I persuade
Πειλάτος Pilate
πεινάω I hunger
πειράομαι I try to
πειρασμός temptation
πειρατής pirate
πέλαγος sea
πέμπτος fifth
πέμπω I send
πένης poor
πέντε five
περί 23
περιβάλλω embrace
Πέρσης a Persian
πεσ- fall
πίπτω I fall
πιστεύω I trust
πίστις belief
πιστός faithful
Πλαταιαί Plataea
πλέθρον 101 feet
πλέω I sail
πλήν but, except
πλήρης full
πλησίον near
πλοῖον ship
πλούσιος rich
ποδ- foot
πόθεν whence
ποῖ; whither?
ποιέω I do, make
ποίημα poem
ποιητής poet
ποῖος of what sort
πολεμέω I wage war
πολέμιος hostile
πόλεμος war
πολιορκέω I besiege
πόλις city
πολίτης citizen
πολλάκις often
πολύ much
πολύς much, *pl.* many
πονηρός bad, evil
πόνος toil, trouble
πορεία journey
πορεύομαι I march, travel
πόσος; how great?
πόσου; at what price?
ποταμός river
πότε; when?
ποτέ once (upon a time)
πότερος; which of two?
ποῦ; where?
πούς foot
πρᾶγμα deed, affair
πράττω I do, fare
πρέσβυς old man, *pl.* envoys
πρίν 89
πρό before
πρόβλημα task, problem
προδίδωμι I betray
πρός 19
προσβάλλω I attack
προσεύχομαι I pray
πρότερος former
προφῆτις prophetess
πρύτανις president
πρῶτος first
πύλη gate
πυνθάνομαι I ascertain
πῦρ fire
πωλέω I sell
πῶς; how?

ῥᾴδιος easy
ῥέω I flow
ῥῆμα word, command
ῥίπτω I hurl
Ῥόδος Rhodes
ῥύομαι I rescue

σαλπιγκτής trumpeter
Σειρήν Siren
σεισμός shaking
σείω I shake
σελήνη moon
σημαίνω I signal, signify
σήμερον to-day
σιγάω I am silent
σῖτος corn, food
σιωπή silence
σκηνή tent
σκότος darkness
σκῶμμα a joke
σμικρός small
σός thy, thine
σοφία wisdom
σόφος wise
σπονδαί truce, treaty
στάδιον furlong

σταθμός stable; halting-place
στερέω I deprive
στολή garment
στόμα mouth
στρατεύω I march
στρατηγός general
στρατιώτης soldier
στρατόπεδον camp
στρατός army
στρέφω turn round
σύ thou
συγγιγνώσκω I pardon
συγκαλέω I call together
συγκοπή fainting fit
συλλαμβάνω I seize
συλλέγω I collect
συμβουλεύομαι I take counsel
συμβουλία advice
σύμμαχος ally
συμφέρει it is expedient
συμφορά disaster
σύμφωνος harmonious
σύν with
συνέπομαι I accompany
συσκευάζω I pack up
σφαῖρα ball, sphere
σφᾶς, σφίσι themselves
σφόδρα exceedingly
σχολαστικός a pedant
σώζω I save, keep
σῶμα body
σωτηρία safety
σώφρων prudent

τάδε these things
τάλαντον talent (£240)
τἄνδον / τὰ ἔνδον what is within
τάξις rank, order
ταράττω I confuse
ταραχή confusion
τάττω I arrange
τάφος burial, tomb
ταχύς swift
τε both, and
τειχίζω I fortify
τεῖχος wall
τέκνον child
τελευτάω I end, die
τέλος end
τέλος *adv.* at last
τετραμοιρία four shares
τέτταρες four
τέχνη skill, art
τῇδε here
τήμερον to-day
τί; why?
τι something, anything
τίθημι I put, place
τίκτω I bring forth, lay
τιμάω I honour
τιμή honour, value
τίμιος valuable
τίς; τί; who? what?
τις, τι someone, a certain
τό the (*neuter*)
τόδε this
τοιοῦτος such
τοξόται police
τοξότης archer
τοσοῦτος so great
τότε in those days, then
τραπ- τρεπ- 82
τράπεζα table
τρέχω I run
τριάκοντα thirty
τριήρης trireme, ship
τριχ- 82
τρόπαιον trophy
τροφή food
τυγχάνω, τυχ- happen
τύραννος ruler, tyrant
τυφλός blind

ὑβρίζω I insult
ὑγιαίνω I am healthy
ὕδωρ water
ὑμεῖς you (*pl.*)
ὑμέτερος your, yours
ὑπέρ *acc.* beyond, *gen.* on behalf
ὑπισχνέομαι I promise
ὕπνος sleep
ὑπό *acc. and dat.* under, *gen.* by, through
ὑπογεῖον basement
ὑποπτεύω I suspect
ὑστεραῖος next, following
ὕστερον later
ὑψηλός high

φαγ- eat
φαίνομαι 80
φάλαγξ phalanx
φανερός plain, clear
φέρω I bear, bring
φεύγω I flee
φημί I assert
φθάνω 80
φθίσις consumption
φιλαδελφία brotherly love

φιλάργυρος miser
φιλέω I love, am wont
φίλιος friendly
φίλος friend
φοβέομαι I fear
φοβέω I terrify
φόβος fear
φοῖνιξ phoenix
φοιτάω I visit
φρέαρ a well
φυγάς exile, fugitive
φύλαξ guard, sentinel
φυλάττω I guard 74
φωνή sound, speech

χαίρω I rejoice
χαλεπός difficult, harsh
χαμεύνιον mat, bed
χαρίζομαι I show favour
χάρις favour, grace
χειμών storm, winter
χείρ hand
χείρων worse
χθές yesterday
χορεύω I dance
χράομαι I use, consult
χρή it is right
χρῄζω I want
χρῆμα thing
χρήματα money
χρήσιμος useful
χρόνος time
χρυσός gold
χρυσοῦς golden
χρυσοτόκος gold-bearing
χρῶμα colour
χώρα land
χωρέω I go
χωρίον a place

ψαύω I touch
ψεύδομαι I tell lies
ψεύδω I deceive
ψηφίζομαι I vote
ψῆφος pebble
ψυχή life, soul

ὦ *interj.* o, oh!
ὧδε thus
ᾠδή ode, song
ᾠήθην *aor. of* οἴομαι
ὠμέγα *last letter*
ὠνέομαι I buy
ᾠόν egg
ὥρα season, hour
ὡς when, as
ὥσπερ just as
ὥστε so that, so as to
ὠφελέω I help

ENGLISH-GREEK INDEX

The numbers refer to pages

abandon καταλείπω 45
able δυνατός 32
about περί 23
absent, I am ἄπειμι
accompany ἕπομαι 32
according to 25
account 12
across = through
advise πείθω, νουθετέω
affair πρᾶγμα
affairs 13 *Note*
after *conj.* ἐπειδή, ὡς
,, *prep.* μετά (*acc.*)
afterwards ἔπειτα
against ἐπί (*acc.*)
Alexander Ἀλέξανδρος
alive ζάω 46
all πᾶς 43
alliance συμμαχία
allow ἐάω 94
allowed, it is 16
ally σύμμαχος
alone μόνος
along κατά 21
already ἤδη
also καί
always ἀεί
ambassador 56
amidst } ἐν
among }
angry, to be 80
announce 54
answer ἀποκρίνομαι 54
any 48
appear 54
appoint 86
archer τοξότης
arm ὅπλον
arm, I ὁπλίζω
army στρατός
around 23
arrange τάττω 37
arrive ἀφικνέομαι 45
art τέχνη
as ὡς
as follows τάδε, ὧδε
as soon as 64
ashamed 80
ask ἐρωτάω 72
ask for αἰτέω

assuredly μήν
at least γε
Athenian Ἀθηναῖος
Athens Ἀθῆναι
attack προσβάλλω (*dat.*)
attempt πειράομαι
avoid φεύγω
away from 16
axe 56

back *adv.* πάλιν
bad κακός
barbarian βάρβαρος
bathe 30
battle μάχη
bear φέρω
beautiful καλός
because ὅτι
become γίγνομαι 45
before *adv.* πρότερον
conj. πρίν 89
prep. πρό (*gen.*)
beg αἰτέω
begin ἄρχομαι 37
behalf 39
behind ὄπισθεν (*gen.*)
believe πιστεύω (*dat.*)
beneath ὑπό
beside παρά 12
besiege πολιορκέω
best, better 62
betray προδίδωμι 76
beyond ὑπέρ 39
bid κελεύω
bird ὄρνις 38
black μέλας 55
body σῶμα 38
bone ὀστέον 50
book βιβλίον
both *adj.* ἄμφω
conj. καί, τε
brains νοῦς 50
brave ἀνδρεῖος
break (treaty) λύω
bribe = persuade with money
bridge γέφυρα
bring φέρω; *fut.* οἴσω
broad εὐρύς
brother ἀδελφός
build οἰκοδομέω

burn καίω 48
bury θάπτω 55
but ἀλλά, δέ
buy ὠνέομαι 50

call καλέω 52
can δύναμαι 32
carry φέρω
cause, to 31
cavalry 61
cease 30
certain τις 48
chain δεσμός
chariot ἅρμα
check παύω
child παῖς, τέκνον
choose 66
circumstances 42
citizen πολίτης
city πόλις 56
claim ἀξιόω
clever σόφος
clever at δεινός
close, I κλείω
come 45, 70
come forward παρελθεῖν
command κελεύω 27
conceal κρύπτω
concern 56
concerning 23
condemn 69
confess ὁμολογέω
confuse ταράττω 37
conquer νικάω
consider νομίζω
consult βουλεύω
convey κομίζω
corn σῖτος
corpse νεκρός
country πατρίς, χώρα
courage ἀρετή
cow βοῦς 60
cowardly δειλός
crocodile κροκόδειλος
cross 70
custom νόμος
Cyrus Κῦρος

damage, I βλάπτω
dance, I χορεύω
dare τολμάω
darkness σκότος
daughter θυγατήρ
dawn ἕως 82
day ἡμέρα

daybreak 43
deceive ψεύδω 38
decide δοκεῖ 49
deed ἔργον
defeat, I νικάω
demand ἀπαιτέω
depart ἀπέρχομαι
deprive ἀποστερίσκω
desire ἐπιθυμία
desire, I ἐπιθυμέω (*gen.*)
desire to, I βούλομαι (*inf.*)
desolate ἔρημος
despise καταφρονέω (*gen.*)
despondent ἀθύμως ἔχειν
destroy διαλύω, διαφθείρω 54
determine βουλεύομαι
die ἀπο-θνήσκω 45; *fut.* -θανοῦμαι
difficult χαλεπός
disaster συμφορά
disease ἡ νόσος
disgraceful αἰσχρός
distant ἀπέχω 16
disturb ταράττω 37
do ποιέω, πράττω
doctor ἰατρός
dog κύων 49
door θύρα
down *adv.* κάτω
prep. κατά 21
draw up τάττω 36
drive ἐλαύνω
dwell ἐνοικέω

each 49
each other ἀλλήλους 28
earth γῆ
earthquake σεισμὸς (γῆς)
easy ῥᾴδιος
educate παιδεύω
Egypt Αἴγυπτος (*fem.*)
eighth ὄγδοος
either...or ἤ...ἤ
elephant ἐλέφας 43
else ἄλλος
encamp στρατοπεδεύω
end τέλος 52
end, I τελέω 52
enemy πολέμιοι
enjoy χαίρω
enough οὕτω(ς)
enslave δουλόω
equal ἴσος
escape ἀποφεύγω
especially μάλιστα
even καί

evening ἑσπερά
events πράγματα
ever ποτέ
every πᾶς
evil πονηρός, κακός
excellent ἀγαθός
except πλήν 43
exist εἰμί
expect ἐλπίζω 31, 38
expedient 56

faithful πιστός
fall πίπτω 57
famous εὐκλεής
far πολύ, πόρρω
fare πράττω 37
farewell χαῖρε
father πατήρ 44
fatherland πατρίς 39
fear φόβος
fear, I φοβέομαι
few ὀλίγοι 62
fight μάχη
fight, I μάχομαι 32
find εὑρίσκω 45
fine *adj.* καλός
finish διαπράττω
fire πῦρ 49
first *adv.* πρῶτον
adj. πρῶτος
fish ἰχθύς 56
flee φεύγω 45
flow ῥέω 48
follow ἕπομαι (*dat.*)
food σῖτος 33
foolish μῶρος
foot πούς 47
for *conj.* γάρ
foreign βάρβαρος
forget 94
forgive 69
former πρότερος
formerly ποτέ
fortify τειχίζω 38
fortunate εὐτυχής 52
free *adj.* ἐλεύθερος
verb λύω
friend φίλος
friendly φίλιος
from ἐκ, ἀπό (*gen.*)
front, in πρό
furlong στάδιον

games οἱ ἀγῶνες
gate πύλη
general στρατηγός
gift δῶρον, give 76
girl κόρη, παῖς
go 45, 70
god θεός
gold χρυσός
golden 50
good ἀγαθός
great μέγας 14
Greece ἡ Ἑλλάς 39
Greek ὁ Ἕλλην 42
adj. Ἑλληνικός
guard ὁ φύλαξ
guard, I φυλάττω
guard against φυλάττομαι
guest ξένος

halt 86
hand χείρ 47
happen γίγνομαι 45; τυγχάνω 80
harbour λιμήν 43
hate μισέω
have ἔχω
head κεφαλή
hear ἀκούω 21, 27, 30
heat θέρος, τό 52
help ὠφελέω
hen ὄρνις 38
herald κῆρυξ 37
hide κρύπτω
highwayman λῃστής, κλέπτης
himself 40
hinder κωλύω
home, homewards οἴκαδε
honey μέλι 63
honour τιμάω
honourable καλός
hope ἐλπίς 39
hope, I ἐλπίζω 27, 31, 38
horn κέρας 82
horse ἵππος
horseman ἱππεύς 60
house οἰκία
household 59
how? πῶς;
however μέντοι
how great? πόσος;
how many? πόσοι;
hungry 46
hurl ῥίπτω
hurt βλάπτω

I ἐγώ
if εἰ, ἐάν
immediately εὐθύς
immortal ἀθάνατος 54
importance 31

impossible ἀδύνατος
in ἐν (*dat.*)
indeed δή, ἄρα
inform 72
injure βλάπτω, ἀδικέω
instead ἀντί 32
insult ὑβρίζω
intend 92
interpreter ἑρμηνεύς
into εἰς, ἐς (*acc.*)
invade εἰσβάλλω εἰς
invite ἐπικαλέομαι
island ἡ νῆσος
islander 13 *Note*

journey ἡ ὁδός
journey, I πορεύομαι
judge κριτής
judge, I κρίνω 54
just δίκαιος
just as ὥσπερ
justice δίκη

keep σῴζω 39
keep quiet ἡσυχάζω
kill ἀποκτείνω 54 (cf. 45)
kind γένος 52
king βασιλεύς 60
knock κρούω
know 68
know how ἐπίσταμαι 32

Lacedaemonian Λακεδαιμόνιος
land χώρα, γῆ
last ἔσχατος
laugh, I γελάω 52
laugh at καταγελάω 52
law νόμος
lawless ἄνομος
lead, I ἄγω, φέρω
learn μανθάνω 45
leave λείπω 45
left *adj.* 83
less ἥττων 62
let 60
let go ἀφίημι 90
letter ἐπιστολή
lie ψεύδομαι 41
lie (down) κατα-κεῖμαι 33
life ψυχή, βίος Ex. 55
lift αἴρω
light φῶς 82
light (a fire) ποιέω, ἅπτω
like *adj.* ὅμοιος 53
like, I βούλομαι (*inf.*); ἥδομαι (*dat.*)

lion λέων 43
listen ἀκούω 21
little ὀλίγος, μικρός
live ζάω, *aor.* ἐβίων
live in = dwell
look on at περιοράω 80
loose λύω
love φιλέω 48

Majesty ὦ βασιλεῦ
make ποιέω 31
make plain δηλόω 50
man ἄνθρωπος, ἀνήρ
many 14
march στρατεύω
market-place ἀγορά
mathematics Ex. 12
may ἔξεστι 16
meanwhile ἐν τούτῳ
messenger ἄγγελος
middle μέσος 14
midst *see* amidst
mind νοῦς 50
mine 32
money 39
month μήν 42
more 49, 62
mortal θνητός
mother μήτηρ 44
mountain ὄρος 52
mourn 30, 48
move κινέω 49
much πολύς 14
music Ex. 12
must 16, 84
my 32
myself 40

name ὄνομα 38
naturally δή, εἰκότως
near πρός (*dat.*), ἐγγύς (*gen.*)
necessary ἐπιτήδειος
necessary, it is δεῖ 16
need δέομαι (*gen.*) 49
neighbour 28
neither *adj.* οὐδέτερος
neither...nor οὔτε...οὔτε
never οὐδέποτε
nevertheless ὅμως
new νέος, καινός
next ὑστεραῖος
night νύξ 37
noble καλός
no longer / no more } οὐκέτι
nor οὔτε, μήτε

nothing οὐδέν
now νῦν *and* 11
number ἀριθμός

obey 30
often πολλάκις
old man γέρων 42
on ἐν (*dat.*)
once ἅπαξ
once, formerly ποτέ
only μόνον
open ἀνοίγνυμι 90
open (letter) λύω
opposition Ex. 48, No. 5
or ἤ
order κελεύω 27
other ἄλλος
others 14, *Note* 3
otherwise ἄλλως, εἰ δὲ μή
ought 16
our, ours 32
ourselves αὐτοί
out of ἐκ (*gen.*)
over ὑπέρ 39
overcome νικάω
owing to 32
ox βοῦς 60

painting ἡ ζωγραφική
pardon 69
pass διάγω
passers by οἱ παρερχόμενοι
penalty 79
people δῆμος
perceive 68
perhaps ἴσως
permit ἐάω 94 *Note*
permitted 56
persuade πείθω
phalanx φάλαγξ 37
pity, I οἰκτείρω
place χωρίον, τόπος
place, I τίθημι 90
plain πεδίον
plainly 80
Plataea Πλαταιαί
pleasant ἡδύς 56
please χαρίζομαι 65
pleasure ἡδονή
poem ποίημα
poet ποιητής
police οἱ τοξόται
poor πένης 52
power δύναμις 56
powerful δυνατός
praise ἐπαινέω 52

pray προσεύχομαι 37
prepare παρασκευάζω 39
present, to be παρεῖναι
preserve σώζω 39
prevent κωλύω
prison δεσμωτήριον
privately ἰδίᾳ
proclaim κηρύττω 37
profitable 56
promise 44
prudent σώφρων 52
punish κολάζω 38
pursue διώκω 30
put τίθημι 90
put to death ἀποκτείνω 54
put up ἀνατείνω

question ἐρωτάω 72
quickly 58
quite πάνυ

race γένος 52
rank τάξις 56
ransom 30
rate 43
rather μᾶλλον
reach = arrive 45
receive δέχομαι 37
reckon νομίζω *and* 31
refuse οὐκ ἐθέλω, οὔ φημι
reign ἐπί 64
rejoice χαίρω 59
remain μένω 54
remain quiet ἡσυχάζω
remember μέμνημαι 32
repent 69
reply = answer
rest, I ἡσυχάζω
rest, the οἱ ἄλλοι
retreat 57
revolt 86
rich πλούσιος
ride ἱππεύω
right(eous) δίκαιος
right-hand 83
river ποταμός
road ἡ ὁδός
robber κλέπτης
ruin 86
rule ἀρχή
rule, I ἄρχω 37
run τρέχω
run away 71
rush 43

safe ἀσφαλής 52

sail πλέω 48
sailor ναύτης
sake 39
same 26
save σώζω 39
saw, I εἶδον 45
say 72
sea θάλαττα
sea-fight ναυμαχία
second δεύτερος
see ὁράω 45, 74
seem δοκέω 49; φαίνομαι 54
seem likely 86
seize ἁρπάζω
self 26
sell πωλέω 50
send πέμπω 39
sentinel φύλαξ
set free λύω
set up ἵστημι
shake σείω 27
shameful αἰσχρός
ship πλοῖον, ναῦς 60
shut (up) (κατα)κλείω 27
signal σημαίνω 54
silence σιωπή
silent, I am σιγάω
silver 50
since ἐπεί and 64
sit κάθημαι 33
slave δοῦλος
sleep ὕπνος
sleep, I καθεύδω
small μικρός 62
snow χιών 52
so (therefore) οὖν
so (thus) οὕτω(ς), ὧδε
so as to 24
so many τοσοῦτοι
soldier στρατιώτης
some 14, 21
someone τις 48
sometimes ἐνίοτε
son παῖς and 13
soon ἤδη, τάχα
sort ποῖος
soul ψυχή
sound φωνή
spear λόγχη
speech λόγος, φωνή
spring ἔαρ 44
stable σταθμός
stand 86
start 43 or αἴρω
stay μένω
still *adv.* ὅμως, ἔτι
stone λίθος
stop παύω (*tr.*); παύομαι (*intr.*)
storm χειμών 43
stranger ξένος
strike κρούω
strong ἰσχυρός
struggle ἀγών, μάχη
such 22
summer θέρος 52
summon καλέω 52
surely 16
surpass νικάω
surprised θαυμάζω 38
suspect ὑποπτεύω
swear ὄμνυμι 90
sweet ἡδύς 56
swift ταχύς 56
swim νέω 48

table τράπεζα
take λαμβάνω 45
take away ἀφαιρέω 66
tall μέγας
tame νικάω
taste γεύομαι (*gen.*)
teach διδάσκω 30
tell (inform) 72
tell (order) κελεύω
temple νεώς 82
terms 20
terrible δεινός
terrify φοβέω
that 20
Thebes Θῆβαι
then (after) ἔπειτα
then (at that time) τότε
there ἐκεῖ, ἐκεῖσε
therefore οὖν, τοίνυν
thereupon ἐνταῦθα δή
thief κλέπτης
think νομίζω, οἴομαι 19
third τρίτος
thirst διψάω 46
this 20
though καίπερ 44
Thracian 51
three τρεῖς, τρία
through διά 12
throw ῥίπτω, βάλλω
thus οὕτως, ὧδε
time χρόνος, καιρός
to-day τήμερον
tomb τάφος
to-morrow ἡ αὔριον

too, also καί
touch ψαύω (*gen.*)
towards πρός 19
train παιδεύω
transact πράττω 36
travel πορεύομαι
traveller ὁδίτης
treaty σπονδαί
tree δένδρον
trireme τριήρης 52
troops = soldiers
trouble ταράττω 36
truce σπονδαί
true ἀληθής 53
trust πιστεύω (*dat.*)
truth τὸ ἀληθές
try, attempt 32, 47
turn 82
twelve δώδεκα
twenty εἴκοσι(ν)
tyrant τύραννος

ugly αἰσχρός
under ὑπό
understand συνίημι 90
unjust ἀδικός
unless εἰ μή
until 64, 89
up ἀνά (*acc.*)
up *adv.* ἄνω
upon ἐν, εἰς
use χράομαι 46
useful χρήσιμος

valuable τίμιος
value τιμή
very μάλα
very *adj.* ὁ αὐτός
victory νίκη
village κώμη
virtue ἀρετή
visit φοιτάω
vote ψηφίζομαι 54

wagon ἅμαξα
wait μένω 54
wall τὸ τεῖχος 52
want δέομαι (*gen.*)
want to βούλομαι (*inf.*)
war πόλεμος
water ὕδωρ 49
way ἡ ὁδός
weak ἀσθενής
weep κλαίω 48
well *noun* φρέαρ 82
well *adv.* εὖ, καλῶς
well-born εὐγενής
what? τί; *adj.* τίς;
whatever 64
when? πότε;
when 64, 66
whenever 64, 66
whether πότερον, εἰ
which ὅς 18
which (of two)? πότερος;
while ἐν ᾧ
white λευκός
who? τίς; 48
who ὅς 18
why? τί;
wife γυνή 37
willing, to be ἐθέλειν
willing(ly) ἑκών 42
win νικάω
wine οἶνος
winter χειμών 43
wisdom σοφία
wise σοφός
wish ἐθέλω, βούλομαι
with μετά (*gen.*)
without ἄνευ (*gen.*)
woman γυνή 37
wonder θαυμάζω
wood (timber) ξύλον
word λόγος
work ἔργον
worse 62
worth, worthy ἄξιος
would that! 66
write γράφω 38

year ἔτος 52, ἐνιαυτός
yet ὅμως
you 18
young νεός
young man νεανίας
your 32
yourself αὐτός
youth νεανίας

Zeus Ζεύς 82

INDEX TO CONSTRUCTIONS AND RULES

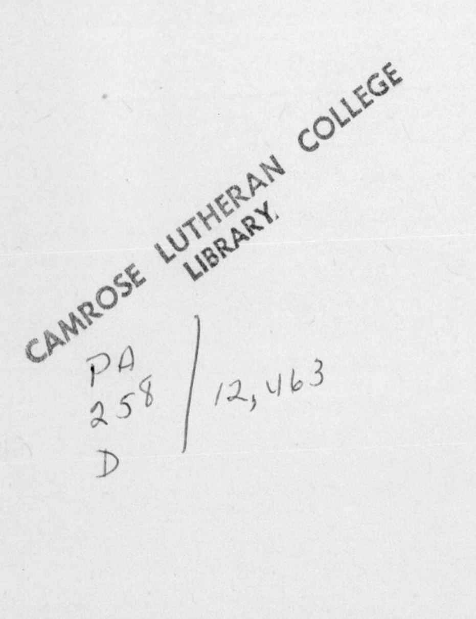